The Chromebook Classroom

THE CHROMEBOOK CLASSROOM

ISBN: 978-0-9977876-0-3

Published by John R. Sowash
Howell, Michigan

Cover and Interior Design by David Schmitke
davideomedia.net

First Printing: June 2016

The Chromebook Classroom

Chromebooks in the K-12 Classroom

JOHN R. SOWASH

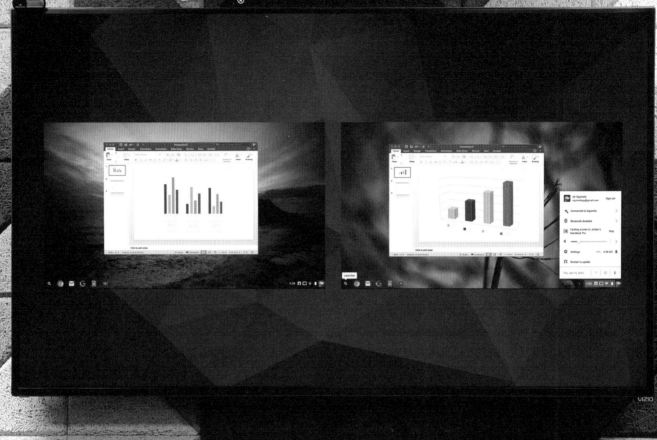

A CHROMEBOOK SCREEN HAS NEVER LOOKED SO **LARGE**.

STARTING AT $14.99 PER CLASSROOM

REFLECTORAPP.COM

GET GOOGLE CERTIFIED

AT THE **GOOGLE CERTIFICATION ACADEMY**

GCA

A high energy, collaborative training program to help educators work toward the Google Certification that is right for them.

Led by a Certified Google Trainer skilled in the classroom use of Google Apps for Education, Chromebooks, and Android Tablets

Gain lifetime access to a vibrant community that will help you become an educational technology leader

Visit gEducator.com/GCA to find a GCA near you!

Don't see an event near you? Host the GCA! No cost to your school and you will receive 3 FREE registrations! Visit gEducator.com/host for details

ACADEMY LEARNING TOPICS

- Gmail Labs
- Automating your inbox

- Drive Add-Ons & Apps
- Google Classrom

- Calendar for lesson planning
- Calendar automation

- Google Play EDU
- Chrome Apps for Classroom

The Google Certification Academy is designed and run by the gEducator team and is not affiliated directly with Google. For information about becoming Google Certified, visit gEducator.com/GCA

DEDICATION

To the thousands of educators who pour their time, passions, and energy into preparing students for the future. Your creativity and dedication are an inspiration.

To my wife - There are very few people in the world who would sign up for the adventure we are on. I'm glad you said yes!

To my five kids, Jonelle, John, Lillian, Ellie, and Caleb - be creative, work hard, and carve out your own paths in life.

ACKNOWLEDGEMENTS

This book was truly a collaborative project. More than 60 individuals contributed ideas, thoughts, revisions, lessons, and suggestions, which made their way into this book. They are listed, by name, in the contributor section.

Writing a book is an expensive dream. I am grateful for the seed funding provided by Air Squirrels (cover sponsor) and Pear Deck (Section Sponsor), which helped get this project off the ground. Go buy their software!

My life-long thanks to my wife and kids for their constant support, encouragement, and patience during long trips and busy weeks of traveling and writing.

Thank you to my editor, Ria Megnin, for sticking through to the end.

Thank you to my designer, David Schmitke, for your mastery of the Material Design visual language.

CONTENTS

Getting Started with Chromebooks

Chromebook Management

Chromebook Lesson Ideas

1 Getting Started with Chromebooks

Introduction

So it's finally happening. Your school is adopting - or considering adopting - Chromebooks for its classrooms. It's time to get up to speed on what this whole "Chrome" thing is about.

When Google emerged as the clear winner of the Internet search engine battle of the late 1990s, it took its mission of organizing the world's data to the next level. Its thousands of talented engineers went to work on hundreds of projects in technology and beyond. Some of those projects did not become commercial successes (i.e. Google Wave, Google Glass) while others become world-changing successes (i.e. Android)

One of the world-changers is Google Chrome, a free Internet browser. First released in 2008, it's become the world's most popular browser, both for desktop computers and smartphones.

What makes Chrome so notable is its ability to work on almost any system - Windows, Mac, Linux, Android, and iOS - along with the opportunity for users to have the same experience across multiple devices. You can open Chrome on your iPad, Android smartphone, Linux computer, Windows PC and your Mac, and your preferences, settings, and content will be identical on all of your devices.

Google's vision for what Chrome could be, however, was limited by the quality of the user's operating system. Older computers often took many minutes to boot up, were plagued by viruses, required a host of manual updates every few weeks, and suffered other serious performance issues. Google decided to bring its trademark simplicity, security, and sense of fun to the operating system itself.

In 2009, the company released a stand-alone operating system called "Chrome OS," designed for users who work (and play!) almost entirely online. This new system provided a super-simple alternative to Microsoft Windows, Apple OS, Linux and other popular operating systems. The entire Chrome OS operating system is built around the Chrome browser, with a few additional functions previously provided by the user's operating system

But the biggest update was yet to come. Having built the browser and the OS, the next step was clear: Google needed Chrome devices.

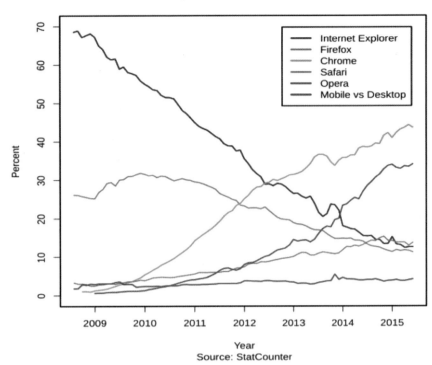

Usage share of web browsers

Source: StatCounter

USAGE OF SHARED WEB BROWSERS

The Chrome browser quickly achieved commercial success, becoming the top browser on the market only four years after its initial launch. Users loved its speed, simplicity and security.

WHAT IS CHROME - VIDEO EDITION

If you are a visual learner, check out a great video explaining ChromeOS.

chrm.tech/sec1-001

Google quietly created an experimental device called the CR-48, designed to test whether a web browser could serve as a fully functional computer. These sample devices were distributed to Google employees and some top user contributors who provided feedback that helped pave the way for the first commercial Chromebook.

Chromebooks are laptop computers built by Google and its hardware partners. They run on the Chrome operating system and are designed for use while connected to the Internet, although they support some offline functionality for a few select Google and third-party applications, including Gmail and Google Drive. Chromebooks are designed to be fully functional computer systems without the clutter, cost, and confusion of traditional devices. This makes them ideal for the classroom!

The first Chromebooks shipped in the summer of 2011 from Samsung and Acer, followed by a variety of others produced by Hewlett Packard, Lenovo, Dell, and other major computer manufacturers. Today, you can purchase more than 50 different versions of a Chromebook, each with a unique combination of design and performance features.

But Chromebook sales started slowly. Many industry analysts criticized Google for its Chrome OS project, saying it was "born to lose" (digitaltrends.com), and "lacks luster and purpose...half baked" (infoworld.com). It even caused Microsoft to launch an ill-conceived ad campaign called "scroogled" which poked fun at the "useless" Chromebook. Despite the critics and the complaints, the Chromebook (and its desktop cousins, Chromebox and Chromebase) continued to spread and improve. Sales were especially strong in the education sector. In the summer of 2015, three years after the launch of the CR-48 test device, Chromebooks became the top-selling educational device in the United States. In early 2016, independent research indicates that 51 percent of devices in the K-12 education world run Chrome OS.

While Google markets Chromebooks and Chromeboxes primarily to first-time computer owners and people seeking a secondary computer for their household, the largest category of Chromebook consumers are schools - like yours. Let's take a look at why.

The CR-48

Google Certified Teachers and Trainers were among those early testers of the CR-48. These test devices were shipped to Google-using educators with no strings attached. Beta testers were asked to use the new device as much as possible and provide feedback on missing features, bugs, and areas of improvement.

Early users quickly realized that a lot of refinement was needed for this new device. Many of the features we take for granted today - the desktop, system shelf, login/out controls, and window switching - were missing from the original Chromebook.

These early users provided mountains of helpful data and feedback, however, and in typical Google fashion, updates and new features were quickly pushed out to this new fleet of web-enabled devices, and incorporated into the generations of Chrome devices to come.

Despite these early challenges, the devices rapidly improved, thanks to bi-weekly operating system updates from Google. To this day, an original CR-48 (now a novelty item) runs the latest version of Chrome OS, despite being the oldest Chrome device in existence!

Chrome Glossary

Google Chrome - A free computer program that is used to connect to the Internet. Google Chrome can be installed on nearly any browser-supporting device (phone, tablet, laptop, etc.)

Chrome OS - An operating system designed for online-focused use that comes pre-installed on Chromebooks and Chromeboxes

Chromebook - A laptop computer running Chrome OS

Chromebox - A desktop configuration running Chrome OS

Chromebase - An all-in-one (monitor and CPU as one unit) computer running Chrome OS

Chromebit - A small HDMI "stick" (similar to a USB drive) running Chrome OS that can turn a television or computer display into a personal computer

Chromecast - A digital media player similar to the Apple TV, Roku, or Amazon Fire TV devices, which can stream content (Audio or Video) from a smartphone, tablet, or other device running the Chrome browser

Chromecast Audio - A digital media player that can turn any speaker into a wireless receiver to stream music from your phone, Chromebook, tablet, etc.

Why Chromebook?

Technology has become an essential part of the school experience. Students use it to learn basic skills, conduct research, and collaborate with classmates. Teachers rely on it for tracking attendance, recording grades, and connecting their classrooms with the larger world. Administrators use technology in every aspect of their jobs! When technology isn't available --or, even worse, is unreliable - it creates significant disruptions in the school day. Reliable, affordable technology that is simple to manage and safe for students is every school leader's dream. The Chromebook makes this dream a reality.

For IT directors, Chromebooks are a dream to manage and monitor due to their web-based Google management console (more on this in Section 2). Content and system setting updates can be set up remotely and "pushed" to an entire district's devices online. These updates can be applied regardless of whether the devices are on or off school grounds, with no popup alerts or forced restarts.

IT IMPACT
Research firm IDC conducted a study on the economic impact of Chromebooks in the K-12 environment and concluded that "Chromebooks reduced the need for additional IT staff to support their deployments, requiring approximately 69% less labor to install and 92% less labor to support than desktop PCs, laptop PCs, or netbooks."
chrm.tech/sec1-004

Unlike Microsoft and Apple's tradition of releasing major new operating systems every few years, Google updates Chrome every two to three weeks with bug fixes and incremental improvements. Significant updates are issued every six weeks. Again, these updates are made automatically, with no interruption to the users, after an extensive review and verification process. This means no forced shift to the latest system releases, and no exposure to the inevitable flood of new system bugs. Not only that, every single Chrome OS update is publicly vetted by early users through Chrome OS's Beta and Development versions, ensuring a smooth transition.

Also, Chromebooks are immune to most of the malware and viruses that plague PCs. Since users can't install program files (.exe) on a Chrome OS system, it's incredibly difficult to corrupt the system. It also means antivirus software and their constant updates aren't needed - although it's still crucial for students to learn and practice safe Internet usage and protect access to their personal information.

Finally, a managed Chromebook can be restricted to a school domain, if desired. That means no personal Gmail accounts could be accessed via those Chromebooks, helping students focus on school-related content.

For teachers, of course, speed is key. Short class periods mean students must be able to pick up a device, power it on, and get to work as quickly as possible. Because personal content is pulled from the same Google "cloud." or Internet-based storage, regardless of which device is used, students don't have to hunt for an assigned laptop. They just pick up a device, log in, and have full access to their personalized content: their saved apps, settings, websites,

Chromebooks Crossing Generations

Special Education teacher Sam Lippert is one of a kind. He served his country in Vietnam, opened a hardware store, managed a restaurant and owns a motel. He then discovered the joy of education. After 24 years as a special education teacher, he's still going strong, even at age 72.

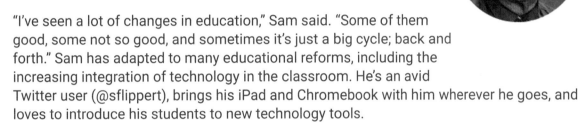

"I've seen a lot of changes in education," Sam said. "Some of them good, some not so good, and sometimes it's just a big cycle; back and forth." Sam has adapted to many educational reforms, including the increasing integration of technology in the classroom. He's an avid Twitter user (@sflippert), brings his iPad and Chromebook with him wherever he goes, and loves to introduce his students to new technology tools.

Sam has seen the rapid advancement of technology in society and education. "I remember students writing on typewriters and when a computer filled a whole room. Today, a computer fits in your pocket."

Sam strongly believes that technology has impacted his students and teaching style in positive ways. "Technology engages students, especially struggling learners, in ways that paper and pencil can't match. It also allows me to differentiate instruction rather than use the same lesson for every student.

"I think classroom technology has the greatest impact on the lowest-performing students. My special ed students have improved their language skills tremendously through tools like Quizlet. I have seen students improve academically by nearly a full letter grade, something I attribute to tools like the Chromebook. Web-based games are engaging them in ways that we can't do with pencil and paper."

Sam's energy and enthusiasm for the teaching profession are inspirational. "I think I've got a few more years in me. I plan on staying in education until I'm at least 75. After that, who knows? I'd be bored if I retired!"

Sam is a great example of a teacher who recognizes and embraces the power of technology in the classroom. For Sam, it's not about Chromebooks or digital flashcards, it's about opening up new opportunities for students who might otherwise slip through the cracks.

documents and creations. All this in less than 10 seconds! Even better, Chromebooks don't get bogged down and slow after a few months of use.

In fact, a Google-commissioned study by research firm IDC found that Chromebooks "increased actual teaching and educational administration time by reducing the time lost in managing desktop PCs, laptop PCs, or netbooks by 82%." That's 82 percent more time to learn about the human body, explore the solar system, or learn how to type, and 82 percent less time spent waiting for devices to update, servers to be upgraded, and new software to be loaded.

Simplicity of use is another huge factor for teachers. Chromebooks are a window to the Web, providing teachers with new learning tools minus the clutter and confusion of traditional technology: Students can quickly pull up a web-based app to explore the human body, start a multimedia project, or log in to Google Classroom to view their latest assignments. Apps can be deployed instantly, and settings can be pre-configured based on classroom needs.

Because Chromebooks are web-based and seamlessly integrate with a user's Google Apps for Education account, students (and educators!) can access their tools, information, projects, and data from their personal and home devices, not just the Chromebook. They can choose to work from their Mac, PC, iPhone, Android tablet, or any other web-enabled device and pick up right where they left off at school. No more messy process of emailing files to a home account or copying files from the school network drive to a USB stick!.

Protection is also a key element of Chromebook design, helping keep students and their computers safe. Chrome OS protects users from phishing and other attempts to intercept login or personal information. It sends personal information using a secure https:// connection, posts warnings when a misleading site tries to open (i.e. www.gooogle.com or www.pintrest.com), and issues an alert if a webpage contains malware.

Chromebooks are even designed to reduce the slowdowns and crashes of other browsers. Most users today like to load many browser tabs at once. Chrome automatically "sandboxes" each tab, so if one crashes or has a loading problem, it doesn't impact the rest of the browser. This feature also helps prevent malicious websites from taking control of the computer.

Finally, teachers in schools that launch a Chromebook initiative get more than just devices. Google offers built-in support for educators, including an extensive collection of resources, trainings, lesson plans, and classroom ideas. (See Resources List at the end of the book for details.)

SIZE MATTERS NOT
Check out the Chromebit - Google's smallest ChromeOS device!
chrm.tech/sec1-005

Chromebooks don't just provide a simple, personal, portable device to support student learning. They provide a flexible computing environment that works anywhere, anytime.

Which Chromebook is right for my classroom?

Chromebooks come in dozens of different versions from many recognizable hardware partners, such as HP, Acer, Lenovo and Dell, with prices ranging from a little over $100 to just over $1,000. The good news is that deciding which Chromebook to buy boils down to only a few factors.

All Chromebooks come equipped with WiFi, Bluetooth, webcams, and integrated microphones.

You don't have to worry about buying expensive batteries. Chromebook battery life is universally outstanding, with all models boasting a minimum six hours of run time and some extending beyond 10 hours of life on a single charge.

Because these devices run on a lightweight OS with cloud storage, concerns about hardware configuration (processor speeds, hard disk space, etc.) become mostly irrelevant. That said, more powerful models do load faster and are able to process high-demand activities, such as video streaming, more easily.

Of course, even Chromebooks have some drawbacks. They are general purpose devices designed to meet most users' everyday needs. Chromebooks are NOT optimized for specialized computing needs, such as high-end graphic design software, proprietary programs, or peripheral support (interactive whiteboards, 3D printers, labware, etc.). Schools will still need to provide access to traditional computers to run specialized software and peripherals. The goal is to have the right devices available when necessary, while having an overall reduction in device and IT support costs.

If you're helping a school district choose which Chromebook to order, here are the key considerations.

Screen Size

Chromebook screen sizes vary widely. Schools with smaller desks or those considering a take-home program should choose models with screens between 10 and 11 inches. This allows students to fit both a Chromebook and a textbook on their desks and to easily transport their computers in their back-packs.

Teachers, on the other hand, do better with a larger, 14-inch screen. This makes updating gradebooks and multi-tasking much easier. Classes that rely on devices for multimedia creation, such as broadcasting or graphic design, also benefit from a larger screen size.

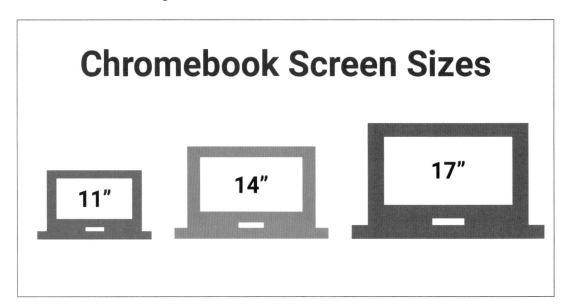

Another important consideration are the requirements for computer-based assessment. Many of the most popular will support testing on Chromebooks, but have requirements related to screen size (usually 11" and above). Be sure to review the hardware requirements for the tests your school has adopted before making purchasing decisions.

Touch Screen Technology

Touch screen technology has become a common feature of all makes of laptop computers. More and more Chromebook manufacturers are offering a touch screen version of their devices, as well.

This may seem like an unnecessary premium feature, but students can use touch screens for many educational activities, including finger writing, tablet-like app interactions, navigation, and learning games. Touch screens also provide benefits for students with special needs, such as improving communication for students with autism or boosting fine motor skills for students with physical disabilities. Meanwhile, the cost of these devices continues to drop, making them nearly the same price as their non-touch screen counterparts. For example, several manufacturers provide touch screen versions of their devices for a small ($30-50) additional fee. For comparison, Chromebooks with touch screens turn out to be less than half as expensive as iPads.

TOUCH SCREEN NEWS:
In the spring of 2016 Google announced that Android Apps would become compatible with Chromebooks, providing another reason to consider touch-screen Chromebooks. Learn more:
chrm.tech/sec1-006

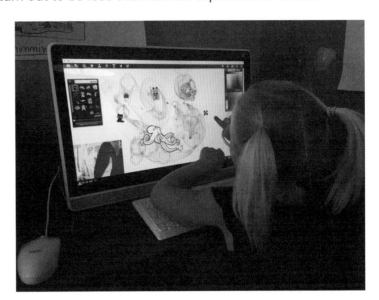

Google has recently indicated that their Android and Chrome platforms will be fully integrated soon. That means schools that purchase touch screen devices will be better prepared to take advantage of upcoming features.

Touch Screen Connection

As a technology integration specialist, I test a lot of new technology and strategies with my own kids. In our house we have an iMac, Chromebooks, a Chromebase, Android Tablets, and iPads. All but 2 of these devices are touch-enabled. When asked which device was their favorite, my kids chose our 21" touch Chromebase from Acer. While the 8 and 6 year old can use a traditional mouse and trackpad, Lillian, age 4, prefers the touchscreen "It's a lot easier to use and random stuff doesn't happen like when I use the mouse. I really like to draw on the computer!"

Is a touch screen an essential feature of a Chromebook? A poll of Chromebook users revealed that 84% of them feel that touch will become an essential device feature.

Durability

Probably the most challenging question for school administrators to consider is the cost/benefit equation of investing in more durable Chromebooks.

FIND OUT

Does a Chromebook bounce when thrown off a balcony?
chrm.tech/sec1-007

Low-end Chromebooks can be purchased for less than $200 per device. These budget devices feature slower processors, lower screen quality and are constructed from lightweight materials, making them more susceptible to physical damage. A low-end Chromebooks may be sufficient for personal, light computing, but will not withstand the rigors of the school environment.

Mid-grade Chromebooks feature shock bumpers around the screen and outside of the device to help the Chromebook survive moderate bumps and falls. Their enclosed keyboards will help protect the device from liquid spills. They also have enclosed screen hinges, addressing one of the most common failure points in laptops. Mid-level Chromebooks offer sufficient, but not extreme, processing power and battery life.

Classroom-optimized Chromebooks offer thicker, heavier construction specifically designed to tolerate more violent drops, dings, bumps, and crashes.

Finally, high-end Chromebooks are customized to provide powerful computing experiences - likely to be most helpful for administrators, teachers, and other staff.

Community Development

The Chrome browser didn't come strictly from Google's development team. It's actually a crowd-sourced effort, built on the best results of The Chromium Project (chrm.tech/sec1-008). The Chromium Project is an open-source developers' version of the Chrome browser and operating system; a constantly evolving "beta" space where users can test and tweak code. They may even see their contributions adopted as a feature of a future Chrome release. The open-source nature of the Chrome platform is an example of how a community of passionate users can build something much better than any individual person or company could.

BUYING A CHROMEBOOK
To find a device reseller near you, visit the official Google For Education Partner directory: chrm.tech/sec1-009

A $149 "consumer" Chromebook tends to last one to three years in a class-room environment. A more expensive "ruggedized" Chromebook costs around $350 and can last three to five years. Which device provides the greatest value? It's a tough question. Chromebooks are on the verge of becoming "disposable" devices. While this trend may not be the best for the environment, it does give school administrators more options around budgeting and device refresh strategies. Being able to regularly cycle new devices into a school gives users access to the latest advancements in memory, processing power, WiFi tech-nologies, etc. At the same time, schools may not have the liquid cash to do a large device purchase every couple of years, especially if the technology is being funded by state and federal grants or local bond proposals. Consider your repurchase funding strategy before choosing the inexpensive or durable route.

Technical "Stuff"

For most other computer purchases, processor speed, memory, and hard disk space play a primary role in device selection. For Chromebooks, these elements are much less important. The average user won't notice a difference between any off-the-shelf Chromebook and a higher-end device.

For "power users," adding additional memory (RAM) to a configuration will provide the biggest performance payoff. Most Chromebooks come standard with 2GB of RAM, which is sufficient, but not excessive. Power users may expe-rience a lag with these devices during daily use. Upgrading to a Chromebook model that offers 4GB of RAM will significantly improve the experience.

Chromebook Purchase Profiles				
PROFILE	**SCREEN SIZE**	**RAM**	**DURABLILITY**	**COST**
The Elementary Student	10-13"	2GB	High	Mid*

Elementary students CAN use Chromebooks! Adding touch capability makes them the perfect device for young learners. A convertible Chromebook can be used as a laptop or as a tablet and is ideal for young learners.

The MS Student	11-13"	2GB	Extra	Low-Mid*

Middle School students can be a bit rough on computers, so pick a device that is designed to take a beating! These devices don't need a lot of extras, keeping their cost low.

The HS Student	11-13"	4GB	Mid	Low-Mid*

High school students do a lot of work with multimedia, so give them a little bit of extra power by selecting a model with at least 4GB of RAM. Students can work on video projects, load a dozen or more tabs, and multitask with ease. Look for a model that offers an enclosed keyboard (limits spill damage) and rugged screen hinges.

The Spreadsheet Ninja	14-17"	4GB	Low	Low*

Spreadsheet Ninjas will appreciate a larger 14"+ screen. The 4GB RAM upgrade will ensure that large spreadsheets load quickly and responsively.

The School Administrator	14-17"	4GB+	Low	Mid-High

Administrators are power users and need a machine that is fast and pleasant to use. Keyboard, trackpad, and screen quality are key. Select a device with an HD display, a powerful processor and at least 4GB of RAM.

The School Secretary	14-17"+	4GB	Low	Mid*

School secretaries can keep track of attendance, email, lunch counts, and more with a large desktop Chromebase. The Chromebase is simply a desktop version of a Chromebook with external keyboard and mouse.

*Pricing Key: Low = $100-$299   Mid = $300-$499   High = $500+

Similar performance gains won't be experienced by purchasing a more powerful processor or additional hard disk space. Almost all processing occurs "in the cloud" rather than on the device, making this upgrade unnecessary. The ability to upgrade your Chromebook disk space (hard drive) is a marketing gimmick that you should avoid! Data is stored in your Google Drive account, not on the device, making disk space upgrades of very little use to the average Chromebook user.

Buying Chromebooks

Chromebooks can be purchased at nearly any technology retailer, however, schools are best served by establishing a relationship with an authorized Chromebook Reseller. Authorized resellers can provide discounted pricing, assistance in selecting the best devices for your needs, technical support, and training to help staff hit the ground running.

Chrome Tour

Let's take a tour of the essential features and elements of the Chrome hardware and operating system.

THE SHELF
The bar that sits at the bottom of the screen. It's the equivalent of the Windows Taskbar or the Dock on Mac computers. The shelf contains the System Tray, App Launcher, and quick-access icons for your most frequently used apps.

The magnifying glass is located in the bottom left corner of your Chromebook. Clicking it gives you access to the App Launcher, where you can browse your installed apps or search the Internet.

Software Features

The first-time setup for a Chromebook is a breeze, one big reason why they're so popular for the classroom. Just log in to any device using either a personal Gmail account or school-provided Google Apps for Education account, and you're ready to learn! No additional setup is necessary, although users can customize many things according to preference.

Once a user has logged in, the Chromebook operating system interface shows just three elements: the desktop, the shelf, and the browser.

By default, several Google Apps are displayed on the shelf, including Chrome, Gmail, Drive and Calendar. Each user can customize which apps appear here. Right click an app on the shelf to display the option to "unpin" and remove it. Right click any of your installed apps to display the option to "pin to shelf." This is helpful for apps that you use on a regular basis - for instance, the Google Dictionary app, Desmos graphing calculator app, or Duolingo app.

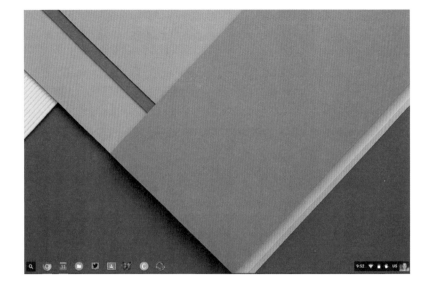

THE DESKTOP:
The display you see when the Chromebook boots up. The desktop displays your open app windows and the wallpaper image of your choice.

THE SYSTEM TRAY:
Located in the bottom right corner of the desktop. It displays the time, current WiFi strength, battery life, and the current user's profile image.

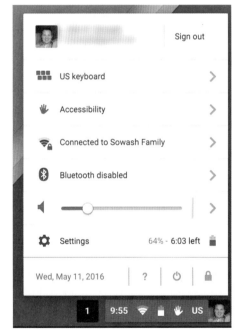

Click on the system tray to view more details: online connections, volume controls, a Help link and a sign-out button. The settings link is also found here. It opens in a Chrome browser window that works even if the computer is offline, and lets users control the wallpaper image and Chromebook theme, choose a default search engine (yes, Bing or Yahoo! can be used from Chrome as well as Google), and options for the keyboard, monitor display, and touchpad or mouse.

The Visual Language of Chrome

Not sure what all of those cryptic symbols mean? Here's a quick rundown!

▦	"The Google Waffle"	Click to access the most popular Google Apps such as Gmail, Drive, and Calendar.
☰	"The Hamburger"	You will see this symbol regularly throughout the Chrome browser. Click to access settings.
⋮	"The Snowman"	All kinds of fun options are hidden under "the snowman." Click to view additional options and configuration choices.
○○○	"The Shish-kabob"	While not as popular as "the snowman", the shish-kabob also gives you access to more settings and commands.

The Chrome Browser

The heart of the Chromebook. This is where the action happens! Click the Chrome "beachball" logo on the shelf to explore, play and work online. The Chrome browser is what makes the Chromebook so great! If you have used Chrome on your Mac, PC, Android phone or iOS device, you will be in familiar territory.

The Chrome Web Store

CHROME WEB STORE
Look for the Chrome Web Store Logo throughout this book to indicate apps that can be installed on your Chromebook.

Your Chromebook is primarily a web browser. That works just fine for checking your email or the weather, but what if you need to edit an image? Create a video? Insert a math equation into a document? The answer is the Chrome Web Store!

The Web Store is a website (https://chrome.google.com/webstore) where you can find all the latest Chrome Apps and Extensions. These allow you to infinitely customize your Chromebook to fit your needs. The Web Store organizes these tools by categories and recommendations, ranging from learning games to personal organizers.

Don't be misled by the "store" name. Nearly all of the apps are absolutely free to download and use. Once downloaded, the apps will be available via the Chrome

App Launcher (the magnifying glass) and any device (Mac, PC, etc.) you sign into that's running the Chrome browser (more on this later!).

The Web Store uses an automated review process to make sure the apps and extensions available are safe. Apps and Extensions that violate user privacy or computer security are removed from the store.

The Google Admin Console gives school administrators and IT directors control over what Apps and Extensions can be installed.

Chrome Apps

Apps are used to accomplish tasks such as learning how to type, editing images and video, creating documents, or exploring the human body. For classroom teachers, Chrome Apps provide an infinite source of new teaching and learning possibilities. And most of them are FREE!

Chrome Extensions

Chrome also supports extensions - simple utilities which increase productivity and simplify common tasks such as checking email or creating a to-do list. Teachers can leverage extensions to increase productivity and manage the classroom.

CHROME EXTENSIONS
Chrome Extensions are utilities that add features into the Chrome Browser. They appear as icons in the top right side of the Chrome browser.

Chrome Themes

The Chrome browser and operating system can be "skinned" with custom themes that alter the color of the operating system and desktop wallpaper. Students and adults alike love to customize their devices by selecting themes.

Hardware Features

KEYBOARD

A Chromebook keyboard is a bit different than the keyboard on other devices you may have used.

While letters, numbers and symbols are set up in the traditional QWERTY layout, many of the peripheral keys and shortcuts you're used to won't be there -- or will have different functions. For instance, instead of using a top row of F labels (F1, F2, F3...), Chromebook features hotkeys along the top and left sides of the keyboard for browser navigation, toggling full screen view, changing screen brightness, and adjusting volume. Google replaced the caps lock key with a special search key providing one-click access to the App Launcher's search bar. But don't worry -- the caps lock function still exists, along with the functions for all the other keys you may have used on PCs or Macs.

If your staff are diehard users of the PgUp, Home, or Command keys, or your classroom absolutely needs a number pad, you can purchase an external keyboard and plug it into your Chromebook using a USB port.

At any time, you can display a map of all Chromebook keyboard shortcuts by pressing Ctrl + Alt + ?.

CHROMEBOOK HOTKEYS	
←	Go to the previous page in your browser history (F1)
→	Go to the next page in your browser history (F2)
↻	Reload your current page (F3)
◹	Maximize your window (F4)
▭	Switcher key. Switch to your next window (F5)
☼	Decrease screen brigthness (F6)
☼	Increase screen brightness (F7)

CHROMEBOOK HOTKEYS

🔇	Mute (F8)
🔉	Decrease the volume (F9)
🔊	Increase the volume (F10)

COMMON WORKAROUNDS FOR MISSING KEYS

Caps Lock	Alt + Search Key
Home	Ctrl + Alt + Up Arrow
End	Ctrl + Alt + Down Arrow
Page Up	Alt + Up Arrow
Page Down	Alt + Down Arrow
Search on Page	Ctrl + F

Chromebook Key Combinations

Alt + Search (magnifying glass key)	Toggles on/off caps lock, for when you need to type out a long acronym or yell at somebody on the Internet.
Ctrl + Shift + L	Locks your screen and requires a password to unlock. A good idea when you step away from your computer.
Ctrl + Alt + ?	Provides an on-screen keyboard that indicates all of the keyboard shortcuts.

Chromebook Key Combinations	
Ctrl + Switcher key	Takes a picture of everything currently on your screen.
Ctrl + Shift + Switcher key	Allows you to select the area you want to capture. *Screen captures are automatically saved in your Downloads folder.
Alt + [Alt +]	Automatically splits your screen. Allows you to view two websites side-by-side. Press once to split 50/50, twice to split 70/30.

TRACK PAD

While it may seem obvious, don't take for granted that people will understand how to use the track pad! Individuals who have used Macbooks and other newer computers will be comfortable with the multi-touch trackpad that is found on most Chromebooks. Students and teachers alike will need to be proficient at single clicking and right clicking. A traditional external mouse can be plugged into a Chromebook using an available USB port. For students, make sure they they are comfortable using the mouse that will be available during computer-based testing.

STANDARD CLICK

Press down on the lower half of the touchpad or lightly tap with one finger.

CLICK AND SWIPE

Drag and drop: Click the item you want to move with one finger. With a second finger, move the item. Release both fingers to drop the item at its new location.

Zoom: Hold one finger down on the touchpad. Swipe left to shrink the page; swipe right to expand it (only works on pages that support zooming, like Google Maps).

RIGHT-CLICK
Click the touchpad with two fingers. (You can also hold down the "Alt" key while performing a single click.)

TWO FINGER SWIPE
Scroll: Place two fingers on the touchpad and move them up and down to scroll vertically, left and right to scroll horizontally.

Swipe: Quickly move two fingers left or right to go backward or forward on web pages or while using apps.

See all open windows: Swipe down with three fingers.

Swipe between tabs: If you have multiple browser tabs open, you can also swipe left and right with three fingers to quickly move between tabs.

Moving to Chrome

Most of today's educators have been using Windows or Apple computers for many years. It can be hard to give up a familiar computer system, but the transition may not be as difficult as you think.

Since the mid-1990s, web browsers - regardless of platform - have taken on a central role in our lives. Many of our most common tasks have shifted online, including banking, emailing, using calendars, organizing photos, using programs, and even creating documents and spreadsheets. Mobile devices such as the iPhone have accustomed us to using single-purpose apps for various tasks. Even on a Mac or Windows computer, it's likely that most of the time, you use a web browser to get your work done.

That means that when compared to the jarring experience of switching between Apple, Microsoft, and other systems in the past, the change to a browser-focused Chromebook can feel easier and much more natural.

Still, it's important to know how Chromebooks compare with the most common systems in our field, and how to adapt your favorite tools and solutions to work with your new device.

Moving from Windows to Chrome

When it comes to basic interface, teachers should feel quite comfortable with the general layout of Chrome OS compared with Windows. Chromebooks follow the Windows model of a taskbar across the bottom of the screen, a "start menu" accessible on the left, and settings on the right.

For many users, the biggest challenge with moving from Windows to Chrome

Getting Staff To Buy In

It's easy to get attached to technology. Teachers used to other systems may put up some resistance when moving toward Chrome. Typically, this is less about Chrome and more about fearing the loss of familiarity and skill with the current platform.

Certified Google Administrator Stephen Gale facilitated a district-wide Chromebook adoption for Westgrand Schools in Kremmling, Colorado.

"It is important to provide staff members with Chromebook training," Stephen says. "Run a Chromebook bootcamp to ensure that everyone is comfortable with the basic operation and functions of the device. That will eliminate a lot of fear."

Another crucial element to support a smooth transition is to conduct an inventory of software, systems, or peripheral devices that may not work on a Chromebook. "We discovered that one of the state reporting tools that we use in Colorado wasn't compatible with the Chrome browser," Stephen says.

Let staff know about these issues in advance, and provide alternate solutions. Stephen and his team, for instance, wrote specific instructions to guide staff in how to submit the required state reports. That helped show that the district had done its research and could assist staff in making the shift to new devices.

Comprehensive staff training and quick responses to compatibility issues will go a long way toward encouraging staff adoption of Chromebooks.

will be the loss of access to Microsoft Office, the longtime de facto choice in office productivity. Chromebooks are, however, able to open Microsoft Office files via Drive, Google's alternative to the Windows-based productivity suite. To make updates and changes to these files, simply convert them into their corresponding Google format (Docs, Sheets, Slides, etc.). These tools are constantly improving, and currently work for the vast majority of users' needs.

Alternatively, users can access Microsoft's new web-based version of Office, Office365. It's compatible with Chromebooks, but features will vary based on the license agreement chosen (education, enterprise, personal, free, etc.).

"Power users" who rely on Microsoft's more advanced but rarely used Windows-based tools, such as Word's suite of mailing and publishing tools or Excel's extensive library of formulas and charts, may want to keep a PC on hand for these tasks.

In addition to MS Office, schools may need to replace a few other Windows-based software titles, such as Windows Movie Maker, Skype, MS Publisher, Adobe Creative Suite (PhotoShop, InDesign, etc.), and keyboarding programs. Many free alternatives, however, can be found in the Chrome Web Store. A few suggested titles can be found below.

Besides software, schools moving from Windows to Chrome may face challenges with peripheral devices: printers, interactive whiteboards, science probes, document cameras, scanners, 3D printers, and most other devices that plug into a USB or VGA port.

NEED TO PRINT?
You will need to setup and configure Google Cloud Print. Learn more:
chrm.tech/sec1-011

Chrome OS does not support driver-based devices, and it is very unlikely that this will change. Google is encouraging all manufacturers to move toward cloud-enabled devices and is leading the way by developing their cloud-based print platform (Google Cloud Print). Until more of these devices are available, you may need a PC to be left in service for certain Windows-based peripheral applications.

Shifting from Windows to ChromeOS also means an adjustment to new concepts in file storage. No longer are files stored locally on a device. They're stored "in the cloud" and are accessible from any web-enabled device simply by logging into your Google Account. This capability is one of the biggest advantages to the ChromeOS platform; however, it can take educators time to adjust to this new reality and fully appreciate its benefits.

The transition to Chromebooks goes best when schools (1) provide training and support for staff as they shift from Office to Drive, and (2) ensure important software and peripherals continue to be supported, either by replacing them with cloud-optimized versions or maintaining old hardware. This support is essential in earning staff buy-in to schoolwide Chromebook deployments, since Chromebooks are not the perfect solution for every situation.

Moving from Mac to Chrome

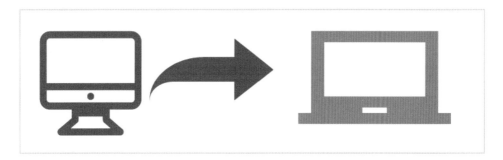

Mac users may find the transition to Chromebooks a bit easier than Windows users, as the physical hardware of a Chromebook is similar to newer Macbooks. Additionally, Mac users are often more accustomed to managing files in multiple formats, from years of having to accommodate the PC-based majority of computer users.

Perhaps the most challenging aspect of transition from Mac to Chromebook will be the loss of the iWork and iLife suite of tools (iMove, Garage Band, iBook Author, etc.). Unfortunately, the conversion of these iWork files to Google Drive is not as simple as the conversion from MS Office to Drive. iWork users will need to save their files into the MS Office format before uploading them to Drive, where they will be converted again into Google Drive format.

Software alternatives

Switching from Windows or Mac requires an adjustment period. It's a good idea to evaluate what software titles are used most frequently by individuals at your school and provide a list of alternatives for them to consider. The following table will help you get started.

Software Alternatives - Windows/Mac to Chrome OS		
Windows/Mac Program	**Chrome OS Alternative**	
Windows Movie Maker	WeVideo ($)	PowToon
iMovie	Animoto Go	Animate ($)
PhotoShop	Pixlr	Sumo Paint
	Picmonkey	

CHROME WEB STORE
To install these Apps, visit the Chrome Web Store and search by name.

Software Alternatives - Windows/Mac to Chrome OS	
Windows/Mac Program	**Chrome OS Alternative**
InDesign Microsoft Publisher iBooks Author	Canva LucidPress BlendSpace Tackk ($) Piktochart EdPuzzle StoryBird Google Drawing Versal BoomWriter
Microsoft Office/iWork	Google Docs, Sheets, Forms, Presentation, Drawing
Skype FaceTime	Google Hangouts
Keyboarding programs	Typing Club/Typing Lessons Typing.com
Mind-mapping software	Google Drawing LucidChart Coggle MindMup Mindmeister
Interactive whiteboard software (SMART, Promethian)	Pear Deck ($) NearPod ($) AirParrot ($)
GarageBand	AuidoTool AudioSauna SoundTrap
iTunes	Google Music Spotify Music Player for Google Drive

$ denotes a tool that is not free or has limited free access

Making The Shift - Chromebook Case Study

Holland High School in west Michigan launched a 1:1 iPad program in 2011. Four years later, in 2015, instead of updating their fleet, they made the switch to Chromebooks.

Technology Integration Specialist Dave Bast said the decision resulted from four key goals the district was trying to accomplish:

- Better integration with Google Apps for Education, which the district was already using.
- Ability to use devices for standardized testing. iPads would require an external physical keyboard for an added cost.
- Budget considerations. The district saved more than $100,000 by purchasing Chromebooks, money which was then allocated for other technology needs.
- Ease of management. App and user management was difficult on the iPads.

"We wanted to make sure that we got technology into the hands of our students," Bast said. "When we looked at how the iPads were being used in the classroom and the direction the district was headed, it made sense for us to switch to Chromebooks."

Bast said the transition from iPads to Chromebooks did involve a learning curve, but by the second semester, staff and students were using them easily and effectively.

Moving from iPads to Chromebooks

The iPad was arguably the first device truly suited for the needs of the K-12 classroom. When it launched in 2010, it quickly became the device of choice for schools due to its long battery life, diverse and high-quality App Store, and simplicity of use. It was, and still is, a marvelous device for schools.

Many districts looking to add additional devices or working to launch a 1:1 program consider both iPads and Chromebooks. It's a real Apples-to-oranges comparison, since these two devices are widely different in purpose and function. For one thing, iPads are designed as touch-screen, single-task devices tied to a single user. Chromebooks are multi-tasking laptops designed for community use in large-scale deployments.

For another, schools that switch from iPads to Chromebooks must replace their favorite iOS apps. The iOS App store is far richer and more developed than the Chrome webstore; however, with access to web-based apps available from the entire Internet, there are few things a Chromebook can't do. In addition, developers of popular iOS apps have begun to port their work to Chrome to access the rapidly growing market of Chromebook users, and Google is working to make Android Apps available to Chromebook users, as well.

Schools must also be prepared to replace the assistive technology features built directly into iPads, such as speech to text, word prediction, and text to speech, with Chrome-friendly tools.

If you're a current iPad user, make a list of your favorite iOS apps and see if a Chrome App version exists. If not, search for a web-based app, which allows you to use a service on a website without needing to download an app. Below is a list of popular iOS education apps and their availability on Chrome OS. As you can see, six are available on Chrome. Two more have numerous online replacement options. That leaves just two apps that Chromebook users will have to do without: an Apple product that can only be used on Apple devices, and one that is not available due to outdated technology (Java).

Software Alternatives - iOS to Chrome OS	
iOS App	Chrome OS alternative
iTunesU	Not available for non-Apple Devices. Alternatives: Google Classroom, Edmondo, Schoology, Moodle, Pear Deck, SeeSaw, Versal
PhotoMath	Not Available for Chrome or web. Alternatives: gMath, GeoGebra, Desmos, WolframAlpha, FastFig
Duolingo	Chrome App Available
Quizlet	Chrome App Available
Peak - Brain Training	Not available for Chrome or Web Alternatives: Lumosity, Twinoo
Mathway	Available via web - mathway.com
Google Classrom	Chrome App Available
Class Dojo	Access via web - classdojo.com
Remind	Chrome App Available
Khan Academy	Chrome App Available
Endless Spanish	Not Available for Chrome. Alternatives: DuoLingo

CHROME WEB STORE
To install these Apps, visit the Chrome Web Store and search by name.

Chromebooks - The Invisible Solution

What makes a Chromebook unique is how invisible it can be - it's designed to be fast, friendly, and simple, so you can focus on what matters. Because Chromebooks are fully web-based, they ensure that all of your content can be accessed from any web-enabled device - your PC at home, the iPhone in your pocket, the Android tablet in your bag, and any other Chromebook you open.

For the classroom, they are perfect. Chromebooks start up in just a few seconds, then provide teachers and students the content they need with minimal barriers and no maintenance downtime. Turn them on, log in, and get started.

Read on for tips on how to deploy a new fleet of Chromebooks and how to manage them in your classroom.

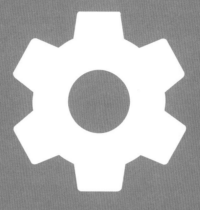

2 Chromebook Management

Introduction

The following section details the deployment and configuration of Chromebooks for K-12 schools. This section will be of particular interest to school administrators and IT staff.

If you are a classroom teacher and are interested in learning the technical nature of how Chromebooks are configured to work so wonderfully in the classroom, read on! If you appreciate the magic, but don't want to know how it's done - skip to Section 3.

For years, schools have dealt with two technology extremes: individually assigned devices or generic "public" devices. Neither of these extremes are ideal:

- Personally assigned computers are modified and customized according to user preference, which means they quickly become bloated with adware and drivers and are frequently missing critical security patches.

- Lab computers are traditionally locked down and wiped after each use to minimize security vulnerability and electronic clutter, leaving them as bland as they are cumbersome.

Over the past few years, schools have greatly increased the number of devices they deploy. We've long since moved beyond most schools having just one or two dedicated labs. It's expected that today's schools will have multiple laptop/tablet carts, dedicated classroom computer sets, or even 1:1 take-home programs.

Google had the distinct advantage of entering this educational technology market late in the game. During the design stage, the engineers behind the Chrome operating system were able to listen to the needs and concerns of teachers, students, and administrators struggling with existing technology for the K-12 environment. All of this information inspired them to build a device that provides maximum flexibility and security, both for end users and for those in charge of making sure the devices work. The end result was the Chromebook, a device that's the balance point between personal and fleet-managed devices. Chromebooks are simultaneously unique for each user, yet easily managed as part of a remote fleet.

Chromebooks have proven to be an effective solution for tiny school districts with part-time IT staff as well as some of the largest school districts in the United States. At a moment's notice, management policies can be remotely pushed to a single device or 10,000 devices. The practical application of these features provides tremendous support for administrators, teachers, and students:

- School principals can have peace of mind, knowing that if a particular feature or tool becomes a distraction, they can adjust devices and user policies to ensure that the student environment is optimized for learning.

- Teachers have access to the best tools the Web has to offer - tools which can be available to students instantly without needing to wait for IT availability or a school break when machines can be collected and re-imaged.

- Students have appropriate freedom to customize their Chrome OS environment to suit their needs and personalities, while being protected from viruses, malware, and inappropriate content.

So how do you take advantage of these features? Read on. This section is for anyone responsible for deploying or managing Chrome OS devices.

First, we'll cover the deployment process for introducing the devices to a full district or school site. Then, we'll share common management approaches for keeping the fleet in good shape. We'll close with guidance for teachers on optimizing Chrome OS devices for your classrooms, from the first to the last day of school.

Setting Up Google Apps for Education

Before attempting to purchase or configure Chrome devices, your school will need to sign up for Google Apps for Education (GAFE). GAFE is provided to schools for no charge and allows a school to use Gmail, Google Calendar, Drive, Sites, and more within their district.

In order to manage, track, and control district-owned devices, schools must purchase a Chrome management license for each device ($30 in 2016). This management license is effective for the life of the device. It can't be transferred or re-used, and it expires when the device it is attached to is no longer in service. This represents one of the best deals in education - you don't want to be without it!

If you purchase your Chromebooks from a Google Partner, the management license will likely be included in the cost of the device. If you purchase Chromebooks from a consumer retail outlet such as Best Buy or Amazon, you will need to connect with a Google Partner and purchase the management license to connect to your devices. A partner directory is available here: chrm.tech/sec2-004

Device Configuration

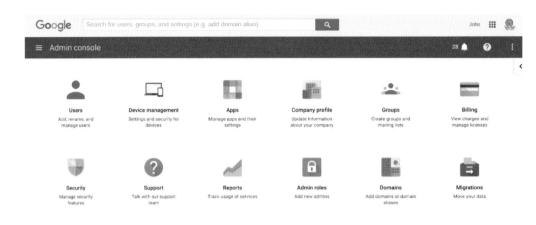

ADMINISTRATIVE SETTINGS
Chromebooks are configured and managed through the Google Apps Management Console available at admin.google.com.

Device Enrollment

The first step in managing your Chromebooks is to enroll (connect) them in your domain. Google maintains a master database of Chrome devices and matches them with the organization that owns them, enforcing that organization's policies and configurations. Any time an enrolled device is powered on, it connects to the database to identify itself. This process ensures that policies are applied no matter where the device is located (home or school) or who is using it. This system is also marvelously effective at recovering lost or stolen devices, even if they are "wiped" in an attempt to hide their identity.

BEFORE YOU BEGIN
Devices must be returned to factory default state (wiped) before they can be enrolled. Device wiping varies by model. Look up the steps here: chrm.tech/sec2-005

The process of enrolling Chromebooks into your domain varies depending on the Chromebook model you have purchased. Review the Google Help article on Chrome enrollment before getting started: chrm.tech/sec2-006

In the GAFE admin console, devices are grouped by OUs. Before enrolling devices, give some thought to how you can best organize your devices for efficient management. Creating an OU for each group of Chromebooks you manage (carts, labs, class sets, etc.) allows you to set permissions according to each group's specific usage needs and purposes. Understand that policies cannot be applied to individual devices - they can only be applied to the OU in which those devices reside.

TRACKING CHROMEBOOKS
Keep track of your Chromebooks by using the Chrome Inventory add-on for Google Sheets: chrm.tech/sec2-007

Understanding Organizational Units

Policies and settings cannot be applied directly to a user or device in the Google Apps administrative console. Instead, policies are applied to "folders" called Organizational Units (OUs). These folders collect similar devices and users into groups and apply a bundle of settings and policies to each OU.

By Organization

▼ sowashventures.com

   ▼ Devices

      Admin Tablets

     ▸ Hill Elementary

      PARCC Testing

     ▸ Park Middle School

     ▸ Wilson High School

The benefit of the OU structure is flexibility, because it allows districts to easily create and adapt multiple sets of policies and configurations based on different groups' needs, then quickly move individual users or devices into OUs as needed.

For example, many schools will create a testing OU configured with policies provided by assessment companies such as Pearson or the Smarter Balanced Assessment Consortium. Most of the time, a district's Chromebooks are in normal config-uration for use by teachers and students. During testing windows, their devices are moved to the testing OU, making them immediately ready to administer assessments. After the testing window closes, the devices are returned to their normal OU, and classroom instruc-tion can resume.

It's time to begin enrolling your devices! Before processing a large number of Chromebooks, it's helpful to determine where the devices will be placed. OUs can be pre-configured before you place devices in them. This supports creating several environments in which the same devices can be used, and moving them into each OU as appropriate. To do so, log in to the GAFE admin console, visit Device management > Chrome management > User settings, and find the "Enrollment Controls" section.

Once enrolled in your domain, devices will appear in the "device management" section of the admin console. Google treats devices as individual units: each one has a unique name (serial number) and can report its current status, location, internal ID numbers, and more. This information can be very helpful when managing Chrome devices remotely or administering a large fleet.

Since Chrome OS automatically pushes out regular, seamless updates, there's no need for IT teams to conduct manual security upgrades or software instal-lations during initial setup. Likewise, cloud-based content and user-based settings eliminate the requirements for a data backup system. Should questions or issues develop, Google provides ongoing support to GAFE schools - look for the "support" section in the admin console.

Device Configuration

After your devices have been enrolled and placed into OUs, configure the hardware and user settings based on your district guidelines and user needs.

The following sections explain the major settings available for Chromebook fleets and who they benefit, along with recommendations for how and when to use them. They are not an exhaustive list covering every possible Chromebook setting, but include the most important configurations to consider.

Hardware Configuration

The Hardware settings regulate who can log in to each Chromebook and how the OU's devices connect to the internet. These settings are typically configured for an entire building or district and operate regardless of who is logged in to a device.

Enrollment & Access

Forcing device re-enrollment allows you to disable a device that's been lost or stolen and improve its chances of recovery. Besides disabling missing devices, you can set up a special message to be displayed on the device after it boots up. Create this message within the Hardware Configuration portion of the management console.

Hardware Configuration > Device Management > Chrome > Device Settings > Enrollment & Access		
Setting	**Description**	**Recommendation**
Forced Re-Enrollment	Re-enrolls your device into your domain if it is wiped.	Force devices to re-enroll into this domain after wiping.
Verified Access	This is the feature that allows Google to identify your school as the owner of a device and apply your configuration.	Enable for content protection.

AN EXAMPLE OF HOW A DISTRICT COULD ORGANIZE ITS DEVICES

Miller Elementary

  Mrs. Jones

  Mr. Smith

Hill Valley High School

  Media Lab

  ELA Cart 1

  ELA Cart 2

  SBAC Testing

Out of Service

Staff Chromebooks

NOTE: MOVING DOMAINS

Chromebooks cannot be moved between domains without wiping and re-enrolling. If your organization has sub-domains, you must choose to enroll your devices into the primary domain OR the sub-domain. Google recommends enrolling devices into the top level domain so that all users, regardless of domain, can access the devices as determined by the admin configuration.

Asset Management

In the event that a device breaks or goes missing, you will need to know its exact location. This can be accomplished by taking advantage of the custom name fields associated with each device. A good naming convention should include a Chromebook's assigned building, classroom, cart, storage space, etc. It's helpful to give devices a physical tag listing this same information.

Sample naming convention:

- Miller Elem. Cart 1 Device 2
- Hill Valley - Lab A Device 6

You can greatly simplify asset management by using a free add-on for Google Sheets called "Chromebook Inventory". This marvelous utility will download a complete listing of your devices from the admin console, allow you to make changes to the asset fields, and re-upload them to the admin console. Chromebook Inventory can also be used to re-assign your devices to different OU's in the event that you move devices from one location to another.

Learn more about Chromebook Inventory: chrm.tech/sec2-007a

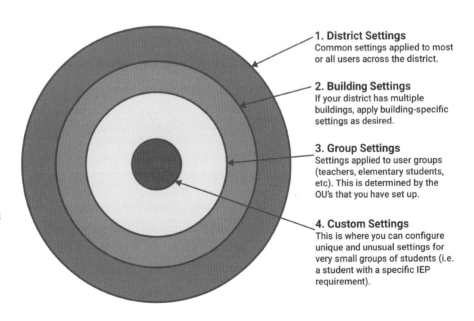

LAYERED POLICIES
Layering your device and user settings will prevent repetitive changes and simplify device management. Configure your most common settings and work your way down, toward specific, special situations.

1. District Settings
Common settings applied to most or all users across the district.

2. Building Settings
If your district has multiple buildings, apply building-specific settings as desired.

3. Group Settings
Settings applied to user groups (teachers, elementary students, etc). This is determined by the OU's that you have set up.

4. Custom Settings
This is where you can configure unique and unusual settings for very small groups of students (i.e. a student with a specific IEP requirement).

Sign-In Settings

Sign-in restrictions control who can access your devices. Individual user policies (discussed later in this section) will be enforced after a user signs in with a school-provided login. If a user is allowed to sign in with a personal account (i.e. Gmail), these user policies will not be applied.

Hardware Configuration > Device Management > Chrome > Device Settings > Sign-In Settings		
Setting	**Description**	**Recommendation**
Guest Mode	Allows a user to browse as a guest (generic user). This will bypass any user settings you have enabled. While this is a helpful feature for devices available for community use, it's not recommended for student devices.	Student devices - Do not allow Staff devices – Allow
Sign-In Restrictions	Configure so that a user must have a school-provided email account to sign in to the device. This ensures that devices are used according to district policy and minimizes theft of devices.	Student devices - ON Staff devices - OFF
Device Reporting	Tracks the current state of the device (past 15 sessions, IP address, system memory, etc.) and the last known user. Helpful in identifying broken, missing, and stolen devices.	Device reporting - ON User reporting - ON
Autocomplete Domain	This setting automatically adds your domain suffix (@myschool.org) to the sign-in screen, shortening the log-in process. This is very helpful for elementary level deployments.	Student devices - ON
Sign-In Screen	Displays the profile photos of the past 12 users who have signed in to the device. Makes the sign-in process easier if only a couple of people use the device. Too many users will clutter the screen.	Shared devices - OFF 1:1 devices - ON Staff devices - ON

Device Update Settings

One of the best features of ChromeOS is its use of automatic, background OS updates. Users receive no warning messages or prompts to restart, download, or install new updates. They simply take place in the background and are applied upon device restart. Google even minimizes bandwidth usage by scattering updates across a fleet, rather than downloading everyone's updates simultaneously. In general, keep the automatic updates settings "on" and allow your devices to migrate to the current version of Chrome.

However, in some instances, you may need to adjust these settings for some or all of your fleet. For instance, some computer-based assessments require a specific version of Chrome for their tests. Testing devices will need to be "frozen" on this version (use the "restrict Chrome version" setting) to prevent updates from being applied during a test session. Refer to the documentation provided by your testing provider (NWEA, PARCC, etc) for specific instructions.

About

Google Chrome OS
The faster, simpler, and more secure computer

[Get help with using Chrome OS]   [Report an issue]

Version 50.0.2661.91 (64-bit)
Platform 7978.66.3 (Official Build) stable-channel samus
Firmware Google_Samus.6300.174.0

✓ Your device is up to date.

[Check for and apply updates]

More info...

Hardware Configuration > Device Management > Chrome > Device Settings > Device Update Settings		
Setting	**Description**	**Recommendation**
Auto Updates	Allows individual Chromebooks to update as soon as a new version of Chrome is launched (every 2-3 weeks).	Most instances - ON
Release Channel	Sets a group of Chromebooks to receive updates from a specific release channel.	Beta channel - 5-10% of fleet Stable channel - standard devices

Kiosk Settings

LEARN MORE
About configuring kiosk mode:
chrm.tech/sec2-008

ChromeOS includes a feature called "Kiosk mode" that allows Chromebooks to be configured for a single purpose, such as filling out a Google Form, taking a standardized test, or accessing a single Chrome application. Kiosk mode prevents people from using the device in normal fashion by locking out all device features except for the app or website it is configured to run. After users finish their session, Kiosk mode returns the device to the starting state.

Kiosk mode can be helpful in many school situations. For instance, it can be used during parent-teacher conferences to allow parents to fill out a parent survey, during a science fair to display a Google presentation to attendees, or in the technology office to let users submit support tickets. Computer-based testing, however, is the most common application for Kiosk mode. Assessments that take advantage of this mode will provide specific instructions on setting up the public sessions that are required.

Hardware Configuration > Device Management > Chrome > Device Settings > Kiosk Settings		
Setting	**Description**	**Recommendation**
Kiosk Settings	Refer to documentation provided by test providers if Kiosk mode is being used for assessment delivery.	No configuration required.

Kiosk Mode in the Classroom

Justin Cowen is a technology integration specialist for Kingsford Schools in Iron Mountain, Michigan. He works closely with teachers, supporting and helping them use technology effectively with their students. One of the favorite web tools being used in Kingsford is Pear Deck, an interactive presentation tool similar to NearPod. During Pear Deck lessons, teachers were concerned that students would wander out of the lesson and into the wilds of the internet. To help address this issue Justin built a custom chrome kiosk app App Using

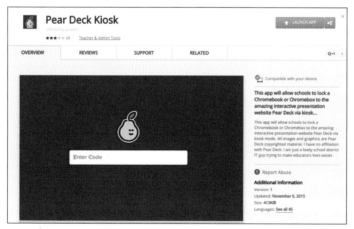

Chrome App Builder (chrm.tech/sec2-009). "Building a simple Kiosk app is a lot easier than people think. It only takes about 20 minutes to build, publish, and deploy a kiosk app to all of our Chromebooks." The Pear Deck kiosk app disables all functions of the Chromebook other than those required by Pear Deck. Furthermore, the browser toolbar is disabled, preventing students from leaving the Pear Deck Session. "Our Pear Deck kiosk app has helped keep students focused and given teachers more confidence in using Chromebooks during whole-class activities."

Public Sessions

The "public session" feature allows community members to quickly check email or browse the web without requiring them to sign in.

To set up and configure a public session, visit: Device Management > Chrome> Public Session Settings. Options include how long a session can last, what apps or extensions can be accessed, and what web pages are automatically loaded.

PUBLIC SESSIONS
Setup a public session to make it easy for parents and community members to check their email or quickly look up information on the web.

Public sessions are great for community spaces such as libraries and media centers, where users need to quickly perform simple tasks such as checking email, filling out a Google Form or looking up information. Configure your public sessions for these situations.

User & Device Reporting

DO IT YOURSELF
Build your own Chrome app! It's easier than you think:
chrm.tech/sec2-010

In most instances, it is helpful to have as much information as possible about the status and health of your Chrome devices. Device state reporting regularly collects status information and reports that information back to the Google Admin console for review. This information can be helpful in identifying broken devices, locating lost devices (by identifying the most recent user), and trouble-shooting WiFi issues.

Hardware Configuration > Device Management > Chrome > Device Settings > User & Device Reporting		
Setting	**Description**	**Recommendation**
Device State Reporting	Provides detailed device status information which can be helpful in maintaining device health.	Enabled

DEVICE REPORTING
Enable device state reporting to monitor the health of your Chromebooks remotely.

Power & Shutdown

The settings in the power and shutdown section are designed for Chromebooks running in Kiosk mode. Chromebooks that are configured for regular use by students or staff should be restarted regularly during normal use, and do not require forced shutdowns.

Hardware Configuration > Device Management > Chrome > Device Settings > Power and Shutdown		
Setting	**Description**	**Recommendation**
Power Management	Determines if an idle device will sleep or shut down.	Allow sleep for general-use devices.
Scheduled Reboot	Forces a device to reboot when running in Kiosk mode.	Leave empty unless configuring devices that are left in Kiosk mode permanently.
Shutdown	Determines how a user can restart the device. Designed to prevent device restarts when running in Kiosk mode.	Leave as default.

Other

When configured, the printer settings listed here will allow users who sign in to Chromebooks (as public users or by using their personal Gmail accounts) to access printers. Most schools provide printer access via individual user rights rather than giving blanket device access. For more on printing from Chromebooks, see the section later in this chapter.

Hardware Configuration > Device Management > Chrome > Device Settings > Other		
Setting	**Description**	**Recommendation**
Cloud Print	Provides printer access to guest users.	Only configure if printer access for guests is desired.
Time Zone	Specifies the time zone for all managed devices.	Check to ensure time is accurate, as this will impact device and app performance.
Mobile Data Roaming	Prevents or allows Chromebooks with data (3G / LTE) connections from roaming.	Configure as desired, if data connections are present.

User Configuration

The previous section focused on device configuration and settings. These settings are applied to devices regardless of user. User Configuration governs what happens after the user logs in to the device, managing what content they can access and what features of Gmail, Drive, Calendar, etc. are available for their use.

It's important to find a balance for these settings between "secure" and "usable." Students (and staff!) are quickly frustrated by devices that are too locked down. That frustration can lead users to turn to unmanaged personal devices, give up on a project - or get creatively destructive and attempt to hack the protections!

That said, maintaining a secure online environment that still meets users' needs isn't easy. It's something districts have wrestled with since the first school servers came online. Instead, try approaching this challenge as an ever-

JOIN THE CONVERSATION
Looking for advice and recommendations? Join the Google+ Chromebook EDU community on: chrm.tech/sec2-011

SAMPLE OU
CONFIGURATION FOR
A SCHOOL DISTRICT
Staff
  Instructional
  Staff
  Support Staff
  Substitutes
Students
  High School
    2017
  Middle School
    2021
  Elementary
  School
    2024
Devices
IT/Administration

evolving, community-led process, rather than an attempt to enforce obsolete rules set by authorities outside the classroom.

Once you've decided how to create an effective online security system, you'll need to set it up and maintain it. Enforcement starts at the IT administration level.

Organizational Unit Structure

Like device settings, user policies are applied to organizational units, not individual users. Before working your way through the user settings in this chapter, consider the user OUs already available. Most schools have different policies for different groups of staff and students. Giving thought to your OU structure early on will significantly improve your ability to manage user and device policies. The more OU layers that are available, the greater the customization possible.

OUs based on anticipated graduation year allow the creation of custom permissions as a student moves through a district. Larger districts with multiple buildings can add in a building layer, allowing building administrations to request custom user permissions and settings based on unique needs or student behavior.

Because settings applied to an OU cascade down to lower levels, most settings can be applied "high up" on the OU tree, making setup relatively simple. Settings can be overridden at lower levels if exceptions need to be applied.

Once your OU structure is configured satisfactorily, you can begin adjusting the ChromeOS user settings. To do so, log in to the admin console and visit Device Management > Chrome > User Settings.

CAUTION: MOBILE
FEATURE
Note that this is
an experimental
feature and that
Google does
not recommend
applying it to all
users at this time.

Mobile

When turned on, this feature allows users to apply their Chrome settings to their Android devices. This could be beneficial if your district is using Android phones or tablets for the classroom.

General

Device Management > Chrome > User Settings > General		
Setting	Description	Recommendation
Custom Avatar	Displays a district-created profile image for public sessions.	Leave empty - or have some fun and upload a unique school icon.

Device Management > Chrome > User Settings > General		
Setting	Description	Recommendation
Wallpaper	Forces a district-assigned wallpaper.	Leave empty or use as an identification and communication tool.
Smart Lock for Chrome	Allows a compatible Android phone to unlock a Chromebook by proximity.	Leave on default setting (OFF).

Pushing Announcements:

Wheatley Park School in Oxford, England builds the school community by pushing out weekly wall-paper updates featuring the Word of the Week and one of the school's community values.

Looking for more ideas? Check out how Franklin Public Schools in Wisconsin uses custom wallpaper to send announcements to students: chrm.tech/sec2-012

Enrollment Controls

Earlier in this chapter, we discussed the initial enrollment of Chrome devices. These settings should be reviewed and adjusted any time you are preparing to enroll new devices into your domain.

Device Management > Chrome > User Settings > Enrollment Controls Overview		
Setting	Description	Recommendation
Device Enrollment	Governs which OU a device will be placed into upon enrollment. "Current location" indicates the top OU of your domain.	Keep device in current location.

Device Management > Chrome > User Settings > Enrollment Controls Overview		
Setting	**Description**	**Recommendation**
Asset Identifier During Enrollment	Allows enrolling users to set custom fields (asset tracking information) upon enrollment.	Do not allow for users in this organization.*
Enrollment Permissions	Indicates which users can enroll and deprovision devices.	Do not allow users in this organization to enroll new or deprovision devices.*

*The above settings are recommended for all users within your domain except those who are responsible for the enrollment and management of Chrome devices. These users should be placed into their own OU (i.e., "IT Department"), and the above settings should be overridden.

When enrolling large numbers of devices it's a good idea to set up a special user account through which you will enroll devices (i.e. chromebooks@ myschool.org). Save time by configuring the Device Enrollment setting to "Place Chrome Device in User Organization" and methodically moving the enrolling user into the OU where a group of devices need to be assigned (i.e. "cart 4"). Once all devices for that OU have been enrolled, move the user to the next OU and repeat.

Apps & Extensions

The Chrome Web Store is a tremendous classroom resource. It's filled with content that can be added to Chromebooks to help teachers and students edit images (e.g. Pixlr), create presentations (e.g. PowToon Presentation), learn how to type (e.g. Type Scout), or edit video projects (e.g. WeVideo). But beware: some Chrome Apps, Extensions, and Wallpapers are not suitable for an educational environment.

The next two sections of the management console gives school administrators the opportunity to review and restrict content that may not be appropriate for student use. These policies will be different for various groups of users in your school (i.e., elementary, middle school, and high school).

Here are three strategies for managing Web Store content:

Allow Everything: No restrictions. This is ideal for your teaching staff, who need to be able to find and review new tools published to the Web Store.

Blacklist Content: Everything is available for users to install except for content that has been specifically blocked by the district.

Whitelist Content: Only content which the district explicitly allows is available for a user to install. Everything else is blocked.

Push Content: Students are unable to install apps or extensions on their own.

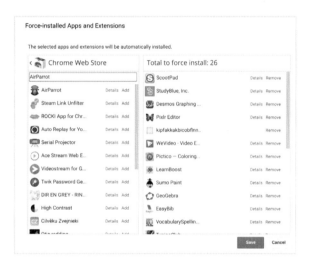

CHROME APP MANAGEMENT
Chrome Apps and extensions can be easily blocked, approved, or pushed out to your entire fleet via the admin console.

Content is deployed only by school administrators and delivered automatically to the device.

Instructional staff and administrators should have unhindered access to all Web Store content so they can discover and test new content for use in the classroom.

Blacklisting and whitelisting can help protect students from content inappropriate for the classroom, but have significant drawbacks. Blacklists will inevitably miss some problematic apps and extensions. Both whitelisting and the push approach make it more difficult for students to discover and access helpful programs.

School administrators and key staff members should gather to discuss the merits of these four strategies, then select a policy that reflects the current values of the school community, the management resources of the IT department, and the needs and concerns of classroom teachers. Remember that online security is an ever-evolving process!

For Schools that are new to Chromebooks and would like a list of the top apps to deploy, check out the Chrome App recommendations in section 1.

Device Management > Chrome > User Settings > Apps and Extensions

Note: The settings below are applied to a specific Organizational Unit. Make sure that you are applying the desired policy to the correct OU!

Setting	Description	Recommendation
Allowed Types of Apps & Extensions	Select or deselect to restrict access to certain types of content entirely. This section ONLY applies to non-ChromeOS devices!	These selections ONLY impact Windows, Mac, and Linux devices.
Apps and Extension Install Sources	Allows content to be installed from outside of the Chrome Web Store. Used in rare situations where a school has a custom web app that must be deployed.	Only use for custom web apps.
Forced Install Apps and Extensions	Installs content to appear the next time a user logs in to the device.	Use to deploy district-wide resources that will be used by all staff or students.
Allow or Block All Apps & Extensions	Sets the district policy for user access to apps and extensions.	Discuss with key stakeholders and choose *allow* or *block* based on district needs and policies.
Allowed Apps and Extensions	Based on the policy set above, this is where you will allow/block apps from the Web Store.	Choose apps that you wish to allow or block.
Pinned Apps and Extensions	A "pinned" app will appear on the Chrome Shelf at the bottom of the screen. This is helpful for frequently used apps such as Gmail, Drive, or Calendar.	Pin apps as desired. Caution: The more apps that are pinned, the more cluttered the Shelf will become.

Chrome Web Store

While this is a separate section in the admin console, it is a continuation of the controls for managing content and access for staff and students.

Device Management > Chrome > User Settings > Chrome Web Store		
Setting	**Description**	**Recommendation**
Chrome Web Store home page	Use this setting to set up a custom Chrome Web Store home page. If users can install content, use the home page to recommend specific app titles to them.	The difference between these two home pages is minor and does not significantly impact the user experience.
Recommended Apps & Extensions	If you chose to use a custom home page, this is where you select the apps that will appear.	Configure as desired.
Chrome Web Store Permissions	Schools can create and deploy custom Chrome Apps. This setting would only apply to the individuals who need permission to publish this content to your domain.	Select for your Admin / IT OU.

Security

The security for both devices and users is a primary concern of school administrators and IT directors. This section determines the policies for user passwords and behind-the-scenes security checks. Rest assured that security is one of the hallmarks of the ChromeOS platform. You've chosen a system with a strong history of security.

Apps and Extensions

Teachers and administrators on Google+ were invited to share how their school handled Chrome Apps and Extensions with students. Of the 60 votes that were cast, nearly half pushed apps to students, nearly a third allowed students to install approved apps only, and just a fifth allowed full access to the Web Store with the exception of known problematic content.

Here are a few sample responses:

Rick Humphrey, Wolf Creek Public Schools, Alberta, Canada - "We have no restriction. We are heavily reliant on our digital citizenship / learning coaches and teacher engagement. To date (5 years), we have only had three incidents of bad extensions installed which involved a tech to figure out. Those times became the teachable moments for all, and not just about apps and extensions they installed."

Chris Tenbarge, Lebanon Special School District, Tennessee - "We recently switched from Blacklist to Whitelist, due to the number of extensions students found that had negative effects on the operation of the machine. That said, 99 percent of the apps used for learning purposes are Admin Push."

Ken Creedon, Warren Township Schools, New Jersey - "We used to allow no restrictions for students, but that quickly got out of hand, and the devices got loaded up with useless apps/extensions. We went to push-only this year. We are a smaller district, so managing it is simple. Staff can request additions or removals for student devices, but once we were set up, we did not see many requests for changes. Faculty access is still wide open, so they can add anything they wish. Works well for us."

Device Management > Chrome > User Settings > Security		
Setting	**Description**	**Recommendation**
Password Manager	When you enter a password into a website using Chrome, you will be asked if you want to save the password for future use. This is the "password manager." When used, saved passwords are encrypted behind the user Google Apps login. As long as the GAFE login is secure, the password manager is secure.	Allow users to configure
Show Password	Displays a button giving users the option to unmask a password and show it in plain text.	Always show
Screen Lock	Locks device after 10 minutes of inactivity on battery power or 30 minutes on AC.	On for all users
Incognito Mode	Incognito mode allows a user to open a web page as an anonymous user. This can be helpful, but can also bypass some content restrictions.	Disallow - Student OUs Allow - Teachers/Staff/OUs
Browser History	Saves browser history across all of a user's devices.	Always save
Force Ephemeral Mode	Ephemeral mode erases all local data upon logout. This setting is not recommended for regular users, but may be helpful for public sessions. Chromebooks automatically manage local data based on available space and the number of users on a device.	Do not erase local user data
Online Revocation Checks	This setting performs an additional security check for secure websites. Leave as default unless requested to change by a software vendor.	Do not perform (default)

Device Management > Chrome > User Settings > Security		
Setting	**Description**	**Recommendation**
Safe Browsing	Protects users from potentially malicious websites (phishing).	Always enable
Malicious Sites	If a user attempts to visit a malicious site, Google will present a warning page indicating danger ahead. This setting allows a user to continue forward, if desired.	Allow user to proceed to a potentially malicious site anyway *Some WiFi login pages will be flagged as malicious sites because they automatically redirect the user to an unrequested page. It is important that the user be able to access the blocked page to sign in to the network.
Geolocation	Allows websites to request user location using IP address triangulation. Right now, this is not a widely used feature on laptop devices. The Google Maps' "current location" feature is one of the few that are operational.	Allow user to configure
Minimum SSL Version	Requires secure website encryption to be at a minimum version. Leave as default unless issues are known to exist.	No policy
Single Sign-On Online Login Frequency	Districts can force users to sign into their Google Account using a third-party platform. Log-in information is sent to Google for authentication resulting in a Single Sign-On (SSO) system. This is an unusual situation, as most districts use Google as their SSO provider.	Leave as default unless using an SSO service
Single Sign-On	Turns on the SSO feature.	Disable unless using an SSO service

Network

The features in this section are advanced network configuration options. In most cases, the default settings can be left as-is.

Device Management > Chrome > User Settings > Network		
Setting	**Description**	**Recommendation**
Proxy Settings	A proxy is a third party that directs web traffic. Most of the time, your devices will connect directly to the internet without a proxy. Your web filtering service may ask you to configure a specific proxy setting to connect to their services. Enterprising students will occasionally find this setting and attempt to configure it to bypass your web filter.	Most circumstances: Do not use a proxy Follow instructions provided by your web filter vendor and configure as directed.

BROWSE SAFE
Chrome's safe browsing feature will warn users about potentially malicious websites.

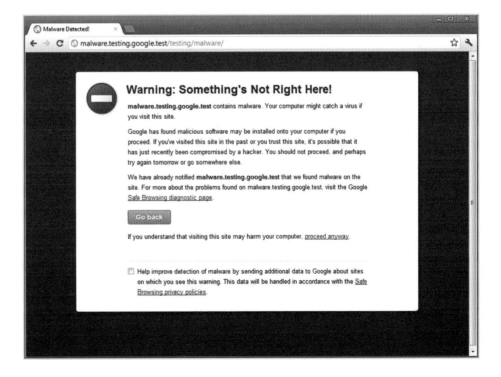

Device Management > Chrome > User Settings > Network		
Setting	Description	Recommendation
SSL Record Splitting	Ensures that secure websites load correctly.	Enable SSL record splitting
Data Compression Proxy	Uses a Google Proxy server that reduces data consumption. Designed for mobile devices on data connections.	Most circumstances: Disable data compression proxy Note: If you have Chromebooks with data connections, you may wish to leave this feature on for those devices. Place these devices in their own OU.

On Startup

Reduce the amount of time it takes students to get their Chromebooks into learning mode by configuring the "On Startup" section to automatically load web pages as soon as a student opens Chrome. Consider using Google Sites to build a student resource page with icons to frequently used resources and teacher pages.

STARTUP PAGES
Elementary teachers will love the ability to auto-open select pages upon login, which helps young students get started faster.

Startup pages ×

M Inbox - jrsowash@gmail.com - Gmail	https://mail.google.com/mail/u/0/#inbox
Google Drive	https://drive.google.com/drive/u/1/folders/0BxBCkVveE609fjVSVGd..
Google Calendar	https://www.google.com/calendar/render?tab=mc
Add a new page	Enter URL...

Use current pages   OK   Cancel

Device Management > Chrome > User Settings > On Startup		
Setting	Description	Recommendation
Home Button	Displays the home button on the browser bar.	Allow user to configure

Device Management > Chrome > User Settings > On Startup		
Setting	Description	Recommendation
Homepage	Determines the page that is loaded when the home button is selected.	Set as desired
Pages to load on startup	Automatically loads specified pages upon login.	Set as desired

Content

Keeping students safe on the web is a big deal. It is our job to ensure that students are protected. While the following Chromebook settings do provide some content restrictions, they are NOT sufficient to safeguard students from inappropriate content. All schools should implement some sort of web filtering solution to prevent access to inappropriate content. Google Apps for Education does not provide a web filter, but many reputable web filter providers can be found that are optimized to work with Chromebooks.

Device Management > Chrome > User Settings > Content		
Setting	Description	Recommendation
Safe Search and Restricted Mode	Determines if Safe Search is used as default for Google Search. Restricted Mode for YouTube is a new feature that only displays content which has been approved as appropriate for the K-12 classroom. Learn more: chrm.tech/sec2-013	Students: Always use Safe Search Staff: No Policy Restricted Mode for YouTube: Configure YouTube via the admin console before making a selection
Screenshot	Allows a user to take screen shots.	Enable Screenshot Note: Some testing providers may require that this feature be turned off when setting up test parameters.
Client Certificates	Allows a district to set up a unique certificate for specialized services.	Configure as requested by software vendors

Device Management > Chrome > User Settings > Content		
Setting	**Description**	**Recommendation**
3D Content	Enables WegGL. This feature will be required in order for certain applications such as BioDigital Human and Google Maps to run.	Always allow display of 3D content
Cookies	Cookies are small, locally stored text files that help websites identify return users and provide a personalized web experience. Cookies can also be used to track user activity. Some websites will not work if cookies are turned off. Sites that use cookies in ways that violate your policies can be blocked.	Recommended: Allow sites to set cookies Determine if a district policy exists for third-party cookies and configure as desired
Third-Party Cookie Blocking	Some websites will allow third parties (frequently advertisers) to install cookies. These cookies can be more problematic, and are more frequently used to track user behavior.	Allow user to configure Users will see a prompt near the URL bar for websites that install third-party cookies. They can then accept or block the cookie.
Images	Determines if images will be displayed on websites.	Show images
JavaScript	JavaScript is a web element frequently used for interactive web features. Some Chrome apps and websites will not work without JavaScript.	Allow user to configure

Device Management > Chrome > User Settings > Content		
Setting	**Description**	**Recommendation**
Notification	This is a new feature for Chrome that allows mobile-style notifications to appear on Chromebooks. Chrome Apps such as Google Tone use this feature, and more websites will take advantage of this service in the future.	Allow user to configure Note: Web notifications are a new area of development and may require additional thought as they become more popular.
Plug-Ins	Plug-Ins such as Flash and the Chrome PDF viewer enable content that Chrome can't natively display.	Allow user to configure
Five settings deal with plug-in permissions and authorization. As a whole, Chrome is moving away from plug-in support in favor of HTML5 and Chrome Extensions. These settings will be left as-is for most districts.		
Pop Ups	Allows users to allow / disable pop up pages.	Allow users to configure Note: Some login screens require pop ups. Allow users to enable pop up pages as needed.
URL Blacklist	Users will be unable to visit any web page listed in this section.	Schools should avoid using this feature except in very unique situations. Use your web filter to restrict access to web page content.
Google Drive Syncing	Allows documents created or edited offline to be synced to Drive on the web. Offline Drive access must be turned on in order for this setting to be effective. Configure in the Drive section of the admin console.	Allow users to decide whether to use Google Drive Sync
Google Drive Syncing over Cellular	Only applies if you have Chromebooks with a data connection.	Leave as default

Google Cloud Print

CLOUD PRINT
To print from a
Chromebook you
will need to set
up Google Cloud
Print for your GAFE
domain. Learn
more: chrm.tech/
sec2-013a

Printing

While we operate in an increasingly digital world, the need to print still exists. Chromebooks do not support traditional device drivers that are required by printers. The lack of driver support increases the speed and security of the Chrome operating system.

Generally speaking, schools can adopt one of three printing strategies:

#1 Don't Print Inform staff and students that printing from Chromebooks is not supported and encourage them to use Google Drive to share files rather than printing them.

#2 Print Kiosk Inform staff and students that if they wish to print they should log in to a Windows or Mac computer. Many schools are setting up "print kiosk" stations in the media center specifically for this purpose.

#3 Cloud Print If printing is an absolute requirement for staff and students, configure Google Cloud Print for your district. This will allow authorized Chrome devices to print wirelessly from anywhere in the world! District administrators have the ability to enable or restrict access to printers by using Google Groups.

Device Management > Chrome > User Settings > Printing		
Setting	**Description**	**Recommendation**
Printing	Enables / disables printing	Configure based on desired printing policy (see above)
Print Preview	Enables / disables the Chrome system print preview window and printing options.	Students: On Staff: Off Staff members who are NOT using Chromebooks may find the Chrome preview window to be restrictive and prefer the native Windows / Mac print options. Disabling Chrome preview will also prevent users from using Cloud Print.

Google Cloud Print Submissions	Enables users to use Google Cloud Print	Enable if Cloud Print has been configured for your district
Google Cloud Print Proxy	Allows users to share their local printers (connected to Mac / PC).	Enable only if you wish users to configure Cloud Print for their classroom printers This setting should be on for the account that you use to set up Cloud Print for the district.
Print Preview Default	Use this setting to define the default printer for users. You can also configure the default printer for public kiosk sessions.	Change to "define default printer" and configure as desired

User Experience

The odds and ends in this section will impact user experience more than many of the other settings we have configured. Choose these settings wisely, and you'll support teachers by helping students gain easy access to important information and use their Chromebooks more effectively.

Device Management > Chrome > User Settings > User Experience		
Setting	**Description**	**Recommendation**
Managed Bookmarks	Allows administrators to push bookmarks to users.	Push important bookmarks out to staff and students. • School webpage • Gradebook • Media center resources • HR resources
Bookmark Bar	Forces the bookmark bar on or off.	Allow user to configure
Enable bookmark editing	Enables / disables the ability for a user to add or edit bookmarks.	Enable

BOOKMARK SPACE SAVER!
For sites that have a custom favicon, save space on the bookmarks bar by leaving the name field empty; the bookmark can be identified by the favicon.

Device Management > Chrome > User Settings > User Experience		
Setting	**Description**	**Recommendation**
Download Location	Sets the location where files will be downloaded. Can be configured to the local downloads folder or Google Drive.	Local Downloads folder Note: In a shared Chromebook environment, students may be confused when downloaded files are not available on different devices.
Spell Check Service	Turns on / off the built-in spell checking service.	Allow user to decide Note: Some teachers may request turning off spell check so that they can test spelling digitally.
Google Translate	Gives users the option of translating web pages into English.	Allow user to configure Note: As with spell check, language teachers may ask for this feature to be disabled for students taking a language course.
Alternate Error Pages	Enables web services that provide suggestions for incorrect URLs.	Allow user to configure
Developer Tools	Developer tools provide deep access to web encoding. Developer tools are available by clicking the Chrome menu and visiting "More Tools."	Always allow

Device Management > Chrome > User Settings > User Experience		
Setting	Description	Recommendation
Form Auto Fill	Google will collect and remember form information such as name, address, phone number, etc. This makes completing forms quick and easy.	Allow user to configure
DNS Pre-fetching	DNS pre-fetching is one of the features that makes Chrome the fastest browser available. Pre-fetching automatically pre-loads all of the links on the page that a user is currently viewing, making them load even faster when clicked.	Always pre-fetch DNS
Multiple Sign-In Access	Google makes it very easy to manage multiple Google accounts. This setting determines whether a user can sign into a second account (i.e. personal Gmail) and what restrictions are enforced. This is a very important setting for schools to consider.	Staff: Unrestricted user access Students: Block multiple sign-in access Note: If you are restricting sign-in for your Chromebooks to district ID, you most likely will want to disable the ability for users to sign in to a personal Gmail account once logged into the device. District policies will NOT be enforced if a user switches to a personal Gmail account.
Unified Desktop	This is a new Beta setting that allows multi-screen support.	No policy

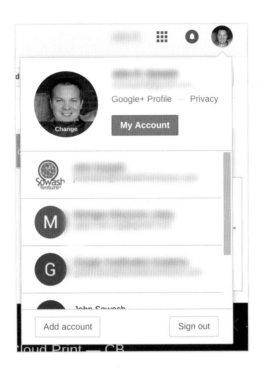

MULTIPLE LOGIN
For districts who are restricting Chromebook sign-in to district email accounts, you will likely also want to disable multiple sign-in access. This will prevent students from also logging into their personal Gmail accounts.

Search Suggest

Search is a key aspect of the Chrome browser. Type anything into the Chrome "Omnibox" to visit a website or perform a search. The following settings determine what options a user has for configuring Chrome search.

Device Management > Chrome > User Settings > Search Suggest		
Setting	**Description**	**Recommendation**
Search Suggest	When search suggest is on, Google will try to complete your search as you type.	Allow user to configure
Omnibox Search Provider	Determines whether a user can change the default search. This setup is quite involved.	Allow user to select configuration

Hardware

The final set of configurations in the Chrome User settings provide additional control over hardware input and outputs. These configurations should be updated with care because they will have a significant impact on users.

Device Management > Chrome > User Settings > Hardware		
Setting	**Description**	**Recommendation**
External storage devices	Determines whether a user can connect an external storage device such as a USB drive or cell phone. For devices with SD card slots, this setting also determines whether SD cards can be read.	Allow external storage devices
Audio Input	Determines whether websites can access the audio port to record sound. This setting does not impact the use of external microphones connected via USB. If disabled, websites such as WeVideo will be unable to record audio via the audio jack.	Enable Input
Audio Output	Determines whether users can play sounds. If disabled, the Chrome device will mute all sounds.	Enable Output
Video Input	This setting determines whether the webcam can be used by websites. Disabling video does NOT disable video for Google Hangouts. To do so, block the Hangouts app from the webstore.	Enable Input
Keyboard	It is possible to reconfigure the top row of keys on a Chromebook to act like function keys on a Windows computer.	No Policy

Protecting Your Chromebooks

Chromebooks aren't immune to damage from bangs, drops and spills - and that's exactly what they'll encounter in a school environment. Before purchasing and deploying Chromebooks, be sure to consider how your district will handle device protection, repair, and replacement.

Breakage rates on Chromebooks will vary based on the type of program offered (take-home programs have the highest breakage rates), the model that is purchased (rugged vs. inexpensive), and the effectiveness of the communication to parents and students regarding their responsibility to protect the device. As a general planning metric, expect 10 to 20 percent breakage for a take-home program.

The low price of Chromebooks allows school administrators to consider several different options for protecting and replacing Chrome devices:

- Buy less expensive Chromebooks, but purchase 10 to 20 percent more than are needed in order to replace broken devices.

- Purchase insurance or protective cases for all devices.

- Buy higher-priced, more rugged devices that will reduce breakage rates.

All three strategies have been implemented by districts with success.

Building a Culture of Student Ownership

The best strategy for protecting your devices is to instill a strong culture of ownership in student and parents. This can be accomplished several different ways:

Assigned Devices

Whether or not schools allow Chromebooks to be taken home, be sure to assign a specific Chromebook to each student. This makes it much easier to track device breakage to specific individuals. It also provides the option of using a meaningful consequence: students who do not treat their devices responsibly will be re-assigned to low-performing devices.

Chromebooks placed in carts should be marked with a unique number or symbol, and students should be required to use the same device on a daily basis. This system works well even when students switch classes. Middle and high school programs, for instance, sometimes distribute Chromebooks during homeroom. Students use and carry the devices for the day, then return them to their homeroom before leaving the school.

Student / Family User Agreement

Every school should adopt a student user agreement which outlines expected behaviors and the purpose of any provided devices. These documents set important expectations and have a significant impact on how school devices are viewed and used (i.e. games and personal entertainment vs. research and study). Create buy-in with a user agreement that empowers users - and protects them. Students, parents, teachers, and administrators should all be involved in the development of district user agreement and acceptable use policies.

Here are some key elements that should be included in your user agreement:

- Acceptable and unacceptable uses for the device
- Privacy expectations (will student data be monitored?)
- Care expectations (is a case required? are stickers okay?)
- Loss, theft, and damage policy
- Consequences of violating the policy (what happens if a student breaks another's device?)

Students should be expected to read and sign this policy before being allowed to use their device.

Parents should also be included in this discussion, as they will play an important role in allowing their student to participate in the program and ensuring that devices are cared for appropriately.

Important elements to share with parents:

- Why do students need a Chromebook? (provide specific learning illustrations)

- Acceptable and unacceptable uses for the device
- Loss, theft, and damage policy
- Financial responsibility and costs (provide options for insurance, if available)

Examples of student and parent device agreements and acceptable use policies are readily available by performing a simple Google search. Here are a few examples to consider.

- Plymouth Canton Schools, Michigan: chrm.tech/sec2-014
- Fair Haven Union High School, Conneticut: chrm.tech/sec2-015
- Mechanicsburg Exempted Village School District, Ohio: chrm.tech/sec2-016

Financial Incentives

Many schools have found it necessary to ask families to make a small financial investment in order to ensure that devices receive proper care. This is most common for take-home programs. This investment can take several forms:

Insurance Plans: Require that parents purchase an insurance plan which covers the device against accidental breakage, theft, and loss. These plans typically start at $30/year with a $30+ deductible. Many schools offer cost sharing for low-income families. Insurance options are covered in more depth later in this section.

Repair Plans: Schools that have the capacity to do device repairs in-house will charge set rates for the work. These repairs tend to be fairly consistent in

DEVICE PROTECTION STRATEGIES
Planning for breakage, repairs, and replacement, is a necessary part of a district wide device deployment. There are several strategies to consider, and each can work effectively with proper planning.

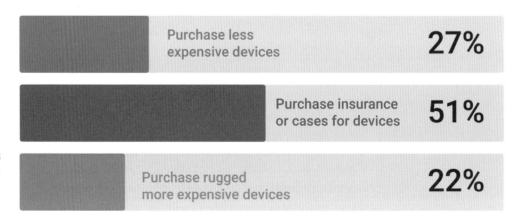

Purchase less expensive devices	**27%**
Purchase insurance or cases for devices	**51%**
Purchase rugged more expensive devices	**22%**

nature (broken screens, missing keys, etc.). Repair prices must be carefully calculated to ensure they cover actual costs.

Lease-to-Own: Some high schools provide families the opportunity to purchase a device over a four-year period. The net cost of this program is very close to purchasing an annual insurance plan ($30-$50/annually). At the end of the four-year period, the student owns the device and can take it off to college.

Device Replacement: In the event that a family elects not to purchase insurance, they may become liable for the full replacement cost of the device in the event that it is lost, stolen, or broken beyond repair. This policy should be outlined in the information provided to parents.

Case Requirement: Cases have been shown to significantly reduce breakage rates on Chromebooks. While some schools provide cases, others require that families purchase a case to protect a student's device. This is sometimes used in conjunction with an insurance policy (i.e. buy a case or buy a plan). Since cases bear the brunt of abuse and cannot always be reused for other students, it's more effective for individual families to purchase them than districts. Furthermore, students can customize their case with stickers and personalization as desired. Read more about cases below.

When considering any of the options above, remember that one of the primary goals of the Chromebook movement is to ensure equal technology access for all. Make sure that family finances won't restrict any student's ability to use Chrome devices.

Chromebook Cases

Cases can help protect devices from the daily rigors of student life. Cases are most popular for schools that offer a 1:1 program, as the rigors of transportation to and from school significantly increase the potential for damage.

Schools that choose not to use cases see damage rates as high as 40 percent, says Mark Zadvinskis, president of protective case maker High Grounds. "Traditional student backpacks offer very little if any padded protection, particularly when the laptop is combined with heavy books, water bottles, chargers, and the many other items found in students' bags. A dedicated case, like the ones manufactured by Higher Ground, provides much-needed protection and will more than pay for itself over the life of the computer."

A wide variety of case solutions are available, including snap-on-shells, padded sleeves, cases with pockets and handles, work-out-of-the-case solutions and more. It's important to choose a case that has been designed for students and schools as opposed to off-the-shelf retail products, which often place style before function and protection. An inexpensive case or neoprene sleeve may look good and help the bottom line, but the lack of protection can end up costing much more.

Another option worth considering is requiring parents to purchase a protective case or damage insurance for a student device. Providing a list of approved cases makes the selection process easier for parents and helps steer the choice to more effective options. Many insurance companies now provide discounts on premiums if an approved case is used, such as those from Higher Grounds.

Schools can also use protective cases as a high-visibility branding opportunity to deter theft and increase the likelihood that a lost device will be returned to the school. Many case manufacturers offer custom embroidery and engraving for large district orders.

CHROMEBOOK CASES
Cases play an important part in protecting devices. These students are using the Datakeeper always-on case from Higher Ground. Learn more: chrm.tech/sec2-017

Chromebook Insurance Plans

Most Chromebook manufacturer warranties last just one year and only cover device defects, not breakage or lost devices. As Chromebooks have grown in popularity, companies have begun offering insurance plans to protect against lost and damaged devices.

Some hardware resellers such as Acer are offering competitive extended warranty programs that expand the length and breadth of the original manufacturer's warranty. Acer's Educare package extends the manufacturer's warranty

for up to three years and expands coverage to include accidental spills and damage. This program does not, however, insure devices against loss or theft.

In addition to manufacturer's warranties and extended warranty plans, device insurance companies have begun to expand their offerings to schools with Chromebook programs. These companies work with both school districts and individual families to cover device loss, theft, and breakage, beyond what a typical manufacturer's plan would cover.

The Worth Ave. Group (http://www.worthavegroup.com) provides a wide range of device insurance options for districts and families. Collinsville Community

Schools in Indiana partnered with The Worth Ave. Group to provide affordable insurance options for families.

Repairing Chromebooks

The large number of Chromebooks being purchased by individual districts and the availability of replacement parts have led some schools to self-insure and repair devices on-site. These districts typically have larger IT departments that can facilitate repairs, or a student-run Chromebook repair course that lets the work happen in-house. Families usually contribute a device insurance fee (usually $20-$50) toward a district-run insurance pool, which is then used to pay for device repairs and replacements.

Mechanicsburg Exempted School District in Ohio decided to set up on-site device repairs when they launched their 1:1 Chromebook program in 2015. The Mechanicsburg technology department consists of a technology director and one assistant, who repair between 10 and 15 of the fleet's 500 devices per week.

"Our device breakage rate in our first year was one percent, which is extremely low," Technology Director Eric Griffith said. "I would attest our success to the culture of ownership that we have developed in Mechanicsburg. Our repairs can be as simple as popping together a separated case bezel or as complete as replacing a logic board or screen. Repairs take an average time of 10 to 15 minutes per incident, and the tech department spends roughly two hours a week on Chromebook repairs. Eventually, our goal is to offer a student repair program to teach students in middle and high school how to service and repair laptops."

Student-led repair programs are gaining popularity due to increasing demands on IT staff and the desire for more STEAM opportunities for students. You can learn more about student repair programs by checking out the Main Township High School (Park Ridge, Illinois) "ChromeDepot" Program at chrm.tech/ sec2-020 and the Niles Township (Skokie, Illinois) Tech Leaders Program at chrm.tech/sec2-021.

Device End of Life

Nothing lasts forever. Chromebooks, like any other computer, will need to fit into a regular progression of deployment, use and maintenance, and retirement. Google provides a master "end of life" document (available here: chrm.tech/sec2-022) to help technology departments plan for the eventual retirement of their device fleet. Google is constantly launching new features and management capabilities which may not be compatible with the hardware capabilities of older devices. This is a normal process for all types of devices and must be anticipated.

Here are the important details from Google's End of Life Policy:

- ChromeOS end-of-life dates are a minimum of five years from the launch of the hardware. These dates may be extended, but will never be contracted.
- After the end-of-life date, automatic updates and management controls are no longer guaranteed by Google.

Every district needs to decide on a strategy for replacing and updating device fleets on a four- to five-year cycle. The range of devices and options, including durability and purchase prices, will give districts many choices to consider.

CONCLUSION

Setting up and configuring a fleet of Chromebooks involves planning and a good investment of time. Compared with other systems, however, the process is both simple and speedy. Chromebooks will also require far less IT support and device downtime over the long term.

If any issues develop, Google Apps for Education customers receive Google Support access. That means a dedicated team of Chromebook experts will be available to help resolve any problems. The Chromebook support team is fast, knowledgeable and friendly (I made lots of calls while writing this book!).

Staying connected with the official Google for Work blog (chrm.tech/sec2-023) will also help districts keep up to date on the latest changes and features to the Admin Console.

Finally, for education environment ideas and support, the Chromebook EDU community on Google+ (chrm.tech/sec2-024) is an invaluable resource. There, you can ask questions and learn more about device management from fellow educators - and share your own ideas.

3 Chromebook Lesson Ideas

Chromebooks in the Classroom

In Section 1, we took a look at the history and purpose of the Chromebook and how it differs from other technology tools, such as the iPad. We also toured the ChromeOS interface and learned its essential features and functions. In Section 2, we dove into the management and configuration of Chromebooks for a K-12 audience, ensuring that our school devices will be safe, secure, and optimized for learning.

In this third and final section, we will explore how the Chromebook can power teaching and learning at all grade levels. We'll look at some of the incredible ways this device can lead students into realms of exploration, discovery, and creativity. We will also tackle some practical considerations - what happens when every student has a device in front of them? How can you effectively guide and direct a group of students who have instant access to a world of information? This final section is about the meeting of technology and pedagogy, and the important ways teachers serve as the guides and facilitators of life-changing learning experiences.

Classroom Management

We all know that classroom management is just as crucial to student learning as lesson design and instruction. Without proper procedures and classroom rules, no technology can be used effectively. Let's add some clear classroom policies regarding the care and use of Chromebooks to your toolkit, including ways to minimize distraction, keep your students safe, and protect district hardware.

In general, your rules should cover three areas:

- protecting and caring for devices
- having students ready at their desks with fully charged, logged-in Chromebooks, and
- making sure device usage is for productive purposes.

Like other classroom procedures, those for Chromebooks must be taught, modelled and reinforced consistently.

To protect and care for devices, consider adopting these safety tips in your classroom:

- When carrying your Chromebook, use two hands at all times.
- When carrying your Chromebook, walk, don't run.
- Chromebooks must be kept flat on your own desk.
- Chromebooks must never be placed on the floor.
- All liquids must stay on the floor.
- Immediately report any physical damage to your Chromebook.
- Don't share your password with anyone.
- Remember to log off when finished.
- Chromebooks should be charged at home or while in the charging cart.

CHROMEBOOK STORAGE
For Chromebooks that are kept in the classroom, make sure you have a safe and secure way to store them. Numbered slots make it easy to see which Chromebooks are missing or in use.

Teachers using Chromebooks in a cart-based setting need to consider how devices will be retrieved, returned, and stored in the Chromebook cart. Here are a few suggestions:

- Don't place the Chromebook cart near the door, as this creates a bottleneck. Place the cart near the back of the room so that students must fully enter before they can pick up a device.
- If the Chromebooks will be used by all of your classes, there is no need to

return them to the cart during the day. Instruct students to log off and leave Chromebooks on the desk for the next class to use.

- It's always a good idea to remind students to log off when finished, even if your devices are configured to log the student out automatically.

- Shared Chromebooks should be numbered, with each student assigned to a particular Chromebook. This provides an accountability measure if any damage occurs to the devices.

Schools that allow students to take their Chromebooks home should consider the following:

- Make sure you have a plan in place for students who forget their Chromebook at home or do not come to school with a fully charged device. Will they be able to borrow a loaner device? Will they be unable to use technology for the day?

- Set up charging stations in the cafeteria or media center for students who didn't come to school with a fully charged device.

- A new fleet of Chromebooks will all look identical. This makes it very easy for students to grab the wrong device. Provide options for customizing and personalizing devices to ensure they are easily identifiable. Cases and covers are good options to consider.

Getting Started

Many teachers are concerned that students won't be able to quickly and successfully log in to their Chromebooks. While this essential activity does require some practice, it's amazing how quickly even our youngest students can master this skill.

To make the sign-in process easier, your IT department can make two small changes:

1. **Turn on domain autocomplete**. This feature (discussed in Section 2) will automatically add the @myschool.com portion of the login, so students only need to type in their user name.

2. **Make passwords memorable**. Google requires that all passwords be a minimum of 8 characters long. Rather than assigning students passwords comprised of random letters and numbers, give them a funny combination of words - i.e. "greensnake" or "purplecow".

Elementary students usually need about a week of practice before they become proficient at signing in to their Chromebooks without assistance. During that

MORE RESOURCES
Additional suggestions for classroom management and device care can be found at the following sites:

Rocklin Unified School District Teacher Expectations and Guidelines: chrm.tech/sec3-001

Chromebook Elementary Rules from Mrs. Schwappach: chrm.tech/sec3-002

Chromebook Rules Infographic: chrm.tech/sec3-003

WHAT'S MY PASSWORD?
Research suggests (chrm.tech/sec3-004) that word-combination passwords are not only easier to remember, they are actually more secure than random letters.

first week, you might find it helpful to provide students with a log-in card to help them remember their usernames and passwords.

DO IT YOURSELF
Use this template to create log-in cards for your students: chrm.tech/sec3-005

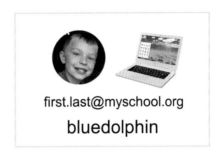

first.last@myschool.org

bluedolphin

Of course, getting students to log in is just the beginning. It's important to help students learn how to shift their attention from the device workspace to the classroom and back. Help them practice the transitions with tools like code phrases. When it's time for students to pause work on their devices to listen to instructions or engage in classroom conversation, try "Heads up, lids down."

Shared Environment

Using Chromebooks in a shared environment such as a dedicated class set or a cart? Here are some specific tips to consider.

Anticipate the next user

Teaching students to anticipate the needs of other users is essential. Stress the importance of maintaining account and device security - students do not want others to be able to access their work or personal information!

4-minute clean up!

Pick a consistent time period for students to log out, safely return devices, and debrief. Consider integrating this clean-up time into a lesson's conclusion.

Classroom "cart captains"

A mass of students trying to return devices to a cart can be a mess! Assign classroom "cart captains," who are responsible for returning their table's Chromebooks to the cart at the end of class and plugging them in.

Monitoring Student Activity

Teachers frequently ask if there is a way to monitor what students are looking at during class. While there are several web-based "lab monitoring" tools available, the most effective and inexpensive tool is your feet! Teachers using Chromebooks should NOT be hidden behind their desk; we need to be engaged and interacting with students throughout the class period. Many teachers have discovered that presenting from a mobile device while walking around the room provides a simple way to keep an eye on what students are looking at on their screens.

That said, it can be helpful to have some technology backup for keeping kids on track! Here are a few lab-monitoring solutions compatible with Chromebooks:

Hapara Highlights - Remotely open tabs, close tabs and even lock screens for a period of time to focus students' browsing activity when required. Learn more: chrm.tech/sec3-006

GoGuardian for Teachers - View student browsing history, monitor on-screen activity, and remotely open and close tabs. Learn more: chrm.tech/sec3-007

gScholar Teacher Dashboard - Allows teachers to view and monitor students' web activity as well as open and close tabs on the students' Chromebooks. Learn more: chrm.tech/sec3-008

Class Dojo - A classroom management system designed for younger students which monitors on-task behavior. Also features a real-time parent tool so parents can track student progress. Learn more: chrm.tech/sec3-009

Another way to minimize student distraction is to engage them in whole-class activities! The following tools allow teachers to present content directly to a student's screen, where they can interact with both the lesson and their classmates:

Pear Deck - An interactive presentation tool which features both content instruction (similar to PowerPoint) and formative assessment. Pear Deck integrates with Google Drive and Google Classroom. Learn more: Peardeck.com

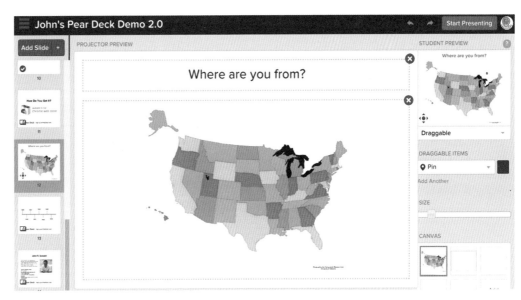

Nearpod - An interactive presentation and formative assessment tool with rich interactive multimedia features. Nearpod also features an extensive user library with thousands of pre-made lessons available for free and purchase. Learn more: Nearpod.com

Kahoot - A wild and crazy game-show-style quiz game. Students compete for high score in the class or to beat their own record. Kahoot is great for in-class review sessions. Learn more: GetKahoot.com

Time Efficiency

Use the following tools to help students boot up, log in, and load lessons quickly!

Get to websites quickly

Every teacher needs a method to get a group of 20 to 35 students to the same website...successfully! Asking them to type in a web URL is NOT the answer! Check out these tools to help your students navigate the internet quickly and efficiently.

Google Tone: <u>chrm.tech/sec3-010</u>

Google Tone can broadcast a website URL to any computer within earshot. This allows instructors to instantly share the correct web address with an entire class, rather than writing it on a board and waiting for each student to access the site individually.

GOOGLE TONE
Getting all of your students to the same website quickly can be challenging. Google Tone is one of many free apps you can use to accomplish this task.

Goo.gl URL shortener: chrm.tech/sec3-011

Use a URL shortener to take a LONG web address and shrink it into a short, manageable one that can easily be typed into a browser URL bar. There are many URL shortener extensions available, but goo.gl is our favorite!

The QR Code Extension: chrm.tech/sec3-012

QR codes are those funny-looking square barcodes. Use this Chrome extension to CREATE your own QR codes, allowing your students to scan and go!

Scan QR: chrm.tech/sec3-013

Have your students use this Chrome App to SCAN the codes you created with The QR Code Extension. This strategy is ideal for early elementary students who are still learning to type.

CHOOSE WHAT WORKS FOR YOU
Review these tools and select one or two to use in your classroom. Consistency is key so that your students know how YOU send out links in your classroom.

Share to Classroom: chrm.tech/sec3-014l

Teachers who belong to Google Classroom can use this handy extension to directly post a website as an announcement, assignment, or question. You can also push the web URL directly to your students so it "magically" pops up on their screens!

Project Management Tools

The following tools and utilities help teachers and students manage their to-do lists, notes, tasks, and more!

Google Keep: chrm.tech/sec3-015

Not sure where to keep that note or idea? Put it in Keep! Google Keep is a flexible, handy tool that organizes your notes, ideas, tasks, favorite sites, and reminders in one place. Install Google Keep on your mobile device so you have all your notes no matter where you are!

Google Spaces: spaces.google.com

Designed for small group collaborating and sharing, Spaces provides a simple way to share links, documents, images, and notes. Spaces is a great choice for group projects.

COMING SOON: GOOGLE SPACES

Project management tools like Google Spaces are great for the classroom. Spaces was launched in the spring of 2016 for consumer Google Accounts and is expected to be available for Google Apps for Education Accounts soon.

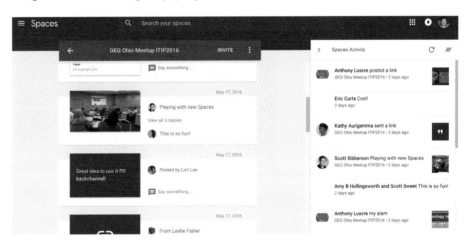

Trello: chrm.tech/sec3-016

Trello is a popular task management tool that is great for groups and teams! Set up your tasks, assign them, and keep track of their status.

Google Tasks: chrm.tech/sec3-018

For users who spend a lot of time in their Gmail inbox, Google Tasks is perfect for keeping track of what you need to do next. Access your to-do list from anywhere using this handy Chrome extension.

Sticky Notes: chrm.tech/sec3-019

If you like Post-It® notes, you'll love Sticky Notes! Stick notes and reminders wherever you need them.

4 Ways to Use Chromebooks in the Classroom

Here are some ways to engage your students in challenging lessons based on whether you have a 1:1 classroom, a cart-based model, or just a handful of devices available.

Large Group Collaboration

LARGE GROUP COLLABORATION
Large group collaborative activities require that all students have access to their own device, which they will use to contribute to a whole-class project.

CHROME WEB STORE
To install these Apps, visit the Chrome Web Store and search by name.

Ideas

Write a class story	Take notes as a class
Pool lab data	Poll the class
Peer edit	Review games
	Offer interactive presentations

Suggested Apps

Google Docs	Padlet
Google Presentation	Kahoot
Google Forms	Pear Deck
Google Sheets	Nearpod
	Socrative

Small Group Projects

SMALL GROUP PROJECTS
Teams of 3 to 5 students work together to research, solve a problem, or create something. Ideally, each student would have their own device. If not, include offline tasks that can be completed while others are working online, to ensure that everyone is contributing.

CHROME WEB STORE
To install these Apps, visit the Chrome Web Store and search by name.

Ideas

Create a website	Create a presentation
Conduct a survey	Make a video
Compile research	

Suggested Apps

Google Docs	Google Drawing
Google Presentation	Google Sites
Google Forms	WeVideo
Google Sheets	PowToon
	LucidPress

Individual Projects

INDIVIDUAL PROJECTS
Students can work individually on assignments from time to time, especially when being introduced to new concepts or ideas. The following suggestions require that students have access to their own, individual devices.

Ideas

Compile research	Practice computer programming
Create a presentation	Solve a math problem
Make a video	Close reading activities
Complete a hyperdoc	
Watch a video	
Create an image or infographic	
Practice typing skills	

Suggested Apps

CHROME WEB STORE
To install these Apps, visit the Chrome Web Store and search by name.

Google Docs	Pic Monkey
Google Presentation	Piktochart
Google Forms	Typing Club
Google Sheets	Typescout
Google Drawing	Gamestar Mechanic
EasyBib	FastFig
Prezi	Desmos
VideoNot.es	Geogebra
Pixlr	

Mini Lessons (Stations)

MINI LESSONS
Mini Lessons involve mixing technology-based instruction with hands-on assignments and class discussion in a series of short activities that students rotate through during a class period.

Elementary teachers are masters at this form of instruction, and middle and high school teachers can use it as well. Here's how: Set up 3 to 5 stations in your classroom with a diverse range of assignments and tasks. Students rotate through the stations in small groups of 5 to 7. This arrangement works even if you only have a couple of Chromebooks in your room, as not all of the stations need to be technology focused.

Ideas

Take a survey	Flip a lesson
Contribute to a document	Ask a question
Watch a video	Explore a manipulative demo
Participate in a debate	

Suggested Apps

Google Docs	ChemReference Tool
Google Forms	Google Earth / maps
YouTube	EdPuzzle
BioDigital Human	CK-12
	Padlet

CHROME WEB STORE
To install these Apps, visit the Chrome Web Store and search by name.

Formative Assessment

Formative assessment is the process of checking student knowledge in order to adjust instruction and meet the learner's needs.The process takes many shapes - asking students to raise their hands in response to a verbal question, have them take quizzes, and send them polls using Google Forms. Formative assessment should be a regular component of all classrooms, allowing for "data-driven instruction." Chromebooks make collecting, analyzing, and responding to student data easier than ever.

Resources

Formative assessment is most effective early in the learning process, such as when a new topic or concept is introduced. It helps identify any misconceptions or learning gaps, so you can adjust your instruction and prepare students for more formal, summative assessments, when they'll be expected to demonstrate what they have learned.

Here are some ideas for incorporating formative assessment into your classroom:

PRE-ASSESSMENT USING GOOGLE FORMS

Create a Google Form that asks your students questions about the topics discussed in class. Assure students that you are not interested in their performance, but in what they they know about the topic at hand. Use the Google Sheet add-on Flubaroo to automatically "grade" student responses. Flubaroo will automatically identify low-scoring questions, so you know what content to focus on.

Sample pre-assessment Form: chrm.tech/sec3-090

Formative assessment doesn't always need to focus on academics. Here's a "mood check-in" form to help you identify students who may not be ready to learn: chrm.tech/sec3-091

DRAWING ACTIVITIES

Many learning concepts have a visual element. To get a sense of what students know, you could ask them to draw a picture, diagram, or conceptual model of key concepts. For instance, they could draw the parts of a plant cell, label major dates on a historical timeline, or fill in a world map.

Tools such as **Ziteboard**, **Sketchpad**, **Google Drawing** and **LucidPress** are all great options for this type of formative assessment. Have students save their initial drawings, then repeat this assignment at the end of the unit. Compare the before-and-after drawings to measure student growth.

INTERACTIVE PRESENTATIONS

As teachers, our hope is to expand the knowledge and understanding of our students. Many times, this requires direct instruction. Lectures *can* be boring, but they don't have to be! **Pear Deck** and **Nearpod** are two excellent tools which combine the best of direct instruction and formative assessment into a seamless activity. Both products allow you to share information while offering interactive quiz questions, drawing activities, web exploration, and even virtual reality! Student responses and interactions are recorded and can be reviewed after the lesson to determine what information needs to be retaught or modified.

REVIEWING CLASSWORK

It's always a good idea to keep track of students as they work on projects and assignments. Reviewing their progress and providing constructive feedback helps them stay focused and on-target. Keeping track of 30+ students can be a difficult task, however! Tools like **Google Classroom**, **Formative**, and ClassKick (classkick.com) make it a lot easier to provide just-in-time feedback.

WHOLE-CLASS REVIEW ACTIVITIES

After providing students with opportunities to learn new information, give them a chance to review and reflect. Here are a few of the many tools that can facilitate individual, group, and whole-class review and reflective activities:

Socrative: Free web-based review and quizzing tool. Features whole-class and small group activities, including the "space race" quiz game!

Kahoot: An immensely popular whole-class review game. Students love it!

Quizlet: Develop a class "set" of information and allow students to review it together or on their own via flash cards and simple review games.

Google Classroom: Use the question feature for exit tickets or class polls.

DATA REVIEW

Don't forget, collecting student data is only the first step. Make sure you take time to review and evaluate student data then adjust your instruction to meet their needs.

Most of the tools mentioned above will provide data in a spreadsheet format. Use the Google Sheets "explore" feature to find trends and threads in your data.

LEARN MORE
Learn more about the Explore features in Sheets:
chrm.tech/sec3-093

Note Taking With A Twist

Taking notes is a pretty standard expectation in the classroom. Students are not, however, naturally good at taking notes! They must be taught and encouraged to take effective notes (writing everything down is NOT an example of good note taking!).

Here are three ways to practice note taking. Using all three strategies at different times is a good idea!

GUIDED NOTES

Provide younger students with an outline with missing portions to fill in. This will speed up their note-taking, allow them to practice their keyboarding skills, and teach them how to identify key points that should be recorded.

Google Classroom is the perfect tool for distributing copies of your guided notes document to an entire class. Select the "make a copy" option when setting up your assignment in Google Classroom.

COLLABORATIVE NOTE TAKING

Take advantage of the collaborative features of Google Docs by having students take notes together. Select two to three students to serve as the official note takers for the class. Give them access to a document and ask them to fill in their notes throughout the class period. You may want to provide them with a guided notes document, as described above.

NOTE TAKING RESOURCES
Guided Notes Example: chrm.tech/sec3-160

Introducing your Students to Collaborative Note Taking:
chrm.tech/sec3-161

Collaborative Note Template:
chrm.tech/sec3-162

Learn more about VideoNot.es:
chrm.tech/sec3-163

While note taking is typically associated with lectures, this method works well for any type of lesson. The goal is to have a summary of what went on during class that day. This document will be a helpful resource for you and your students as you prepare for your next major assessment - especially students who were absent from class.

Using the same document for an entire unit keeps all of the class notes and resources in one location. Provide all students with comment-only access to the document, so they can review the class notes and leave comments or revisions suggestions for anything that was left out.

VIDEO NOTES

For teachers who incorporate lots of video or adopt the flipped classroom method of instruction, the free Chrome app **VideoNot.es** will be an awesome resource. Students use this app to take notes while watching a video, and attach their typed notes to the timeline of the video for easy review. VideoNot.es integrates with Google Drive, allowing finished notes to be easily shared and reviewed.

Special Challenges

Language instruction, math instruction, and special education require a bit of extra thought and planning when using a Chromebook. Professionals will benefit from having access to some content-specific tools.

Math On A Chromebook

The language of math can be challenging for anyone using a traditional keyboard. Mathematical symbols and formulas are difficult to enter quickly and accurately. The following tools and suggestions will help math teachers navigate electronic lessons with students:

Google Docs Equation Editor - The equation editor in Docs has been improved from the first generation. If you haven't used it in a while, give it another chance; you might be surprised by how much easier it is! To access, create a new Google Document and open Insert > Equation.

g(math) - This math tool is an add-on for Google Docs, Sheets, and Forms. You can use g(math) to insert an equation, statistical display, or math expression. g(math) provides multiple input methods including a traditional equation editor, handwriting recognition (great for touchscreen Chromebooks!), and even voice input! Install g(math): chrm.tech/sec3-020

G(MATH) ADD-ON
High School Math teacher John McGowen developed g(math) for his math courses. You can use it for free by installing the g(math) add-on for Docs, Sheets, or Forms. Learn more: chrm.tech/sec3-020a

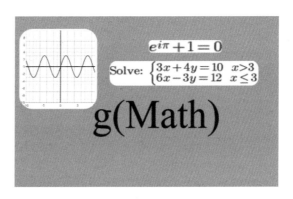

Desmos - Turn your Chromebook into a powerful graphing calculator with Desmos, a free Chrome app. Developed by two California math teachers, Desmos is a potential replacement for TI-84 style calculators. Save, share, and show your work! Install Desmos at chrm.tech/sec3-021

GeoGebra - Teaching geometry? Give GeoGebra a try! This free Chrome App allows students to explore mathematical concepts and share their work. GeoGebra also integrates with Google Classroom. Install Geogebra: chrm.tech/sec3-022

FastFig - This handy Chrome app is like Google Docs...for math! Teachers can use FastFig to easily create a math worksheet to share with students. Students can use the built-in calculation tools to answer questions and show their work. Install FastFig: chrm.tech/sec3-023

Language On A Chromebook

Teaching a foreign language? Here are some great options to make your work easier. Some are optimal for classes of beginning language learners, others are recommended for advanced user only.

LANGUAGE INPUT
Add a secondary language to your Chromebook to easily switch between different input languages.

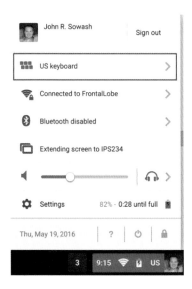

Easy Accents - For students who are just starting to learn a second language, the Easy Accents Add-on for Google Docs provides quick access to special characters not found on an English keyboard. Install Easy Accents Add-on: chrm.tech/sec3-024

ChromeOS Language Settings - Chromebooks are configured with a default language, but also support multiple other languages. Visit the Chrome settings and search for "Language." Once you've configured a secondary language, you can easily switch between primary and secondary options via the system tray in the bottom right corner of the screen.

Google Account Language Settings - Your Google account also has a default language. This impacts your use of Google Search and products such as Gmail, Calendar and Drive. You can configure a secondary language with the Google language settings at: chrm.tech/sec3-025

Google Input Tools - To get one-click access to your Google Account language settings, install the Google Input Tool Chrome extension. Install Google Input Tools: chrm.tech/sec3-026

The impact of technology on world language instruction

Language instructors have a love / hate relationship with technology. The access to tremendous study resources comes with ample opportunities for students to cheat and take shortcuts - such as online translators. But Wheatley Davis, a high school Spanish and technology teacher in Warren, Michigan, says the benefits far outweigh the drawbacks.

By teaching students proper use of technology and directing them to better tools, such as WordReference or Quizlet, Wheatley says she can create a much more effective learning environment.

"Including digital content in my course makes it more relevant to my students," she says, "and often gives them the ability to practice independently in a way they enjoy."

Beyond boosting the learning process, technology shifts "foreign" language away from dry textbooks and distant lands into an immersion experience, right in the classroom. Today's new instructional tools "allow my students to become active language learners through access to authentic listening materials, reading content, movies/videos, music, and cultural connections," Wheatley says.

For ideas on integrating tech with language instruction and more, connect with Wheatley on Twitter (@WheatleyDavis) or on her website, www.wheatleydavis.com.

Special Education

Technology can be a huge boost for students with disabilities, helping them more fully access the curriculum and engage with the classroom. Chromebooks provide multiple supports to accommodate a range of needs. Some of these supports are built into the Chrome OS, while others require third-party extensions or applications to be installed on a student's Google account.

ChromeOS Accessibility Tools - To use the built-in accessibility controls, visit the Chrome settings and search for "accessibility." Most of the options are designed to help people with visual impairments.

Search for the following tools by name in the Chrome Web Store:

Text-to-Speech Extensions - Several Chrome extensions can be used to read a web page's content aloud to a user. It's a good idea to have more than one of these extensions available, as some will not work on particular webpages.

- SpeakIT
- Select and Speak
- Voice Instead

Speech-to-Text - Students who have a difficult time using the keyboard or trouble with spelling can benefit from speaking text aloud, instead.

- Voice Typing for Google Docs - This service is a default feature of Google Docs. To access it, open a Google Document and visit Tools > Voice Typing
- Voice Note II
- Dictanote

Reading Fluency - Students who struggle with reading fluency or dyslexia may benefit from the following tools.

- **OpenDyslexic -** Shifts web fonts into the OpenDyslexic font, which is designed to make text more easily readable for people with dyslexia.
- **Dyslexia Word Highlighter -** Identifies words as the mouse pointer passes over them.
- **BeeLine Reader -** Adds a subtle color gradient to on-screen text to make it easier to distinguish between lines. Also supports OpenDyslexic font.
- **MagicScroll Web Reader -** Changes web pages into a book-style format that eliminates the confusion of traditional webpage scrolling motions.

- **ReadLine -** Presents text one word at a time. Designed for speed-reading, it reduces distractions while reading web pages.
- **Readability -** Strips web pages of distracting ads, videos, and other multimedia content. Works great in combination with some of the above tools, such as Voice Instead and ReadLine.

Texthelp - The technology company Texthelp provides assistive learning solutions for students and teachers. Their products, which assist with reading fluency, language instruction, writing, and math fluency, are especially helpful for students with multiple needs. Texthelp is one of the rare providers to offer product support by both email and phone. Learn more at chrm.tech/sec3-028

- **Read & Write For Google Chrome (RW4GC)** - Texthelp's flagship product provides comprehensive reading and writing support in Google Docs and reading support with websites.
- **Snapverter** - This add-on extends the functionality of RW4GC by converting images of text and inaccessible pdfs into documents that can be read with the text-to-speech tool.

Offline Access

Chromebooks can, if necessary, be set up to work without internet access. Configuring offline access is only recommended in situations where students are assigned to a single device. Offline access is not a realistic option in a shared device situation. Remember, Chromebooks are designed to run on the web. Using them offline disables some of the best features of the device.

Configure the following services to use your Chromebook offline.

Google Drive

To enable Drive offline, each individual user must log in to their Drive account while still online, visit Drive settings (look for the gear icon in the top right corner of the screen), and check the box that says "Enable offline access." A copy of your entire Google Drive account will be downloaded locally on your Chromebook making it available for use when you go offline. These files will be automatically synced when you reconnect to a network.

The following table will help you understand the offline capabilities of Google Drive.

Offline Capabilities of Google Drive			
	View	Edit	Notes
My Drive Folders	Yes	No	Cannot create new folders or organize existing files offline
Docs	Yes	Yes	Collaborative, research and other web-enabled features will not be available
Sheets	Yes	Yes	Collaborative, research and other web-enabled features will not be available
Presentation	Yes	Yes	Collaborative, research and other web-enabled features will not be available
Forms	No	No	Collaborative, research and other web-enabled features will not be available
Drawing	Yes	No	Collaborative, research and other web-enabled features will not be available
Non-Drive Files	Yes	No	Files that can be viewed in Drive will be viewable. Other files (.psd, .indd, etc) will be unavailable

OFFLINE DRIVE ACCESS
Offline access for Drive must be enabled by the user and is device-specific.

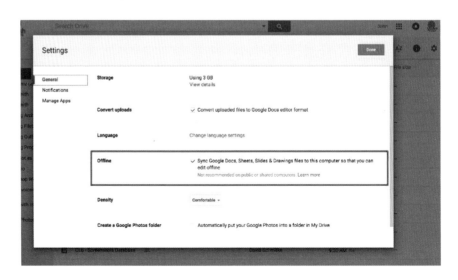

Note that changes made to a collaborative document that is edited by others while you are offline will result in sync issues which will need to be addressed individually.

Gmail

LEARN MORE
about using Drive offline:
chrm.tech/sec3-029

The Gmail Offline Chrome app must be installed and configured prior to going offline. This special app will allow you to read, compose, archive, label, and star messages. Once you reconnect to a network, your changes will sync back to Google.

Calendar

OFFLINE MAIL
Install the Gmail Offline Chrome App:
chrm.tech/sec3-030

Google Calendar is limited to view-only in offline mode. To enable, open Google Calendar and visit settings (gear icon) and find the "offline" option. You will be able to select which of your calendars are available offline once the initial sync is completed.

Keep

Google Keep is handy for taking quick notes, reminders, and creating a to-do list. Google Keep automatically syncs an offline copy of your notes; no configuration required!

Advancements in Adaptive Technology

During his 25-year career, Jeff Crockett has seen many changes in the field of adaptive technology and special education.

"The rapid pace of 21st century technology has provided both new opportunities and challenges for special education," Jeff says. "Never before have we had access to low-cost devices and tools to help with text to speech, speech to text and word prediction."

While these new tools provide terrific help for his students, Jeff wisely notes that they can only be used to their fullest potential if a school provides careful planning, training, and support.

"Proficient use of these new supports by both teachers and students depends on quality professional development and a sustained professional learning community," he says, "which need to be planned for and fiscally supported. Students with disabilities should expect an increasingly accessible curriculum and an improved opportunity to meet curriculum standards due to these advancements in technology."

Offline Video

Making video available for offline watching is possible, but will require some extra work and planning. Currently, subscribing to YouTube Red ($9.99/month) is the only way to make YouTube videos available offline. Videos saved to Google Drive are not supported offline either. To make videos available for offline viewing, users must manually download them as files in the Chromebook's local storage, where they will be available for play.

Teachers who use the flipped classroom method of instruction will want to make videos available for their students via Drive and remind them to download those videos to the local storage when an internet connection is available. This method works most effectively for students who don't have internet access at home and are able to plan in advance. Remind classes that unexpected internet outages can occur even if they have access from home!

OFFLINE ACCESS
The lightning bolt icon in the Chrome Web Store indicates apps that are offline compatible.

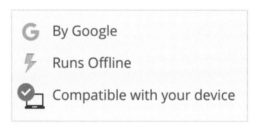

Other Offline Tools

Some apps in the Chrome Web Store also feature offline capabilities. Look for the lightning bolt on the app description, which indicates offline compatibility. You can also browse all of the offline-capable apps from this link: chrm.tech/sec3-031

Here are a few helpful titles to install (search by name in the Chrome Web Store):

- **DuoLingo** - language learning
- **The 50 States Free** - geography game
- **Pattern Shapes by the Math Learning Company** - geometry game
- **Gliffy Diagrams** - graphic organizer tool
- **Fraction Wall by Visnos** - math fraction practice
- **Polarr Photo Editor 3** - photo editing

- **Pixlr Touch Up** - photo editing
- **Robust Audio Recorder** - records and saves audio
- **Evernote** - document and reminder tool
- **Skill Builder Spelling Beta by Kaiserapps** - spelling word practice
- **Scientific Calculator**
- **Webcam Toy** - photobooth app

OFFLINE CONFIGURATION
Some of these apps require that they be opened and configured at least once prior to being run in offline mode.

Lesson Ideas for Elementary Students

Students of any age can use a Chromebook. The lessons in this section start with the basics (learning how to use a computer) and become more complex as the chapter progresses, matching the progression of skills that an elementary student should develop throughout the school year. Modify the complexity and details of each assignment based on the ability of your students.

Many of the suggested lessons in this section work well as station-based or enrichment activities and do not require that you have a device for every student in your classroom. Students of all ages enjoy working in pairs on Chromebooks, so don't be afraid to put them in small groups.

Getting Started with Computers

About the Project

Young students will need to learn basic computer skills and knowledge in order to use their Chromebooks well. Lead them in the following activities early in the school year, so they can get comfortable using a Chromebook sooner.

Mouse / Trackpad Activities

Navigating with a mouse or trackpad can be difficult for little hands and fingers. Here are some exercises to help build mouse skills:

- **Mousercise**: chrm.tech/sec3-032
- **MiniMouse Games**: chrm.tech/sec3-033
- **Voxel Building Blogs**: chrm.tech/sec3-034
- **Letter Tracing**: chrm.tech/sec3-035

LESSON RESOURCES
Printable Keyboard Template: chrm.tech/sec3-036

Computer Terminology Bingo: chrm.tech/sec3-037

Printable Bingo Card: chrm.tech/sec3-038

Practice Logging In

Even young students should have their own account to log into, not a generic class account. They CAN do it! Give students time to practice and master this skill. Turn it into a game! Keep track of how long it takes the entire class to log in. Write down the times in a visible location. Each day, challenge your class to see if they can beat the previous class log-in record.

Understanding Computers

One of the best tech lessons we can give our students is the chance to build their technology vocabulary and knowledge of how computers work. Here are

a few activities that build technology literacy in young students:

PAPER COMPUTERS

Before allowing students to use a "real" computer, have them build a computer out of paper and identify important parts such as the trackpad, screen, power port, USB port, search key, etc.

COMPUTER BINGO

As a class, make a list of all of the important parts of a computer. Write each of the parts on a slip of paper and put them into a bowl or basket. Have students fill out a bingo card with some of their favorite components, then play BINGO as a class.

SIMON SAYS - CHROMEBOOK EDITION!

Provide each student with a Chromebook. Give students simple tasks to perform: open a new tab, log out, open the Learn To Count app, close the lid, etc.

Mary Hankins, K-5 Technology Teacher, Owosso Public Schools, Michigan.

Typing for Elementary Students

About the Project

Much like handwriting, typing is an essential life skill that students must learn as quickly as possible. Chromebooks are compatible with many free resources and programs that can help students improve their typing speed and proficiency. Typing practice is a great enrichment and reward activity once other work has been completed. Regular practice is key.

Typing Resources

Dance Mat Typing: chrm.tech/sec3-039

- Includes good verbal instructions (headphones will be helpful!)
- No login or configuration required, however, progress is not saved or tracked.

Typing.com: www.typing.com/tutor/

- Teacher must set up a class and enroll students to view progress and collect data.
- Free version is ad-supported

Typing Club: chrm.tech/sec3-041

- Features a Chrome App
- Saves student progress, but data is not shared with the teacher.

Typing Lessons: chrm.tech/sec3-042

- Get started fast! No setup or login required for practice.

- Log in using Google to track progress.
- Features a fun multiplayer game

Typing Practice Games

- Typing Aliens Attack: chrm.tech/sec3-043
- Typing Games: chrm.tech/sec3-044

Station Activities for the Elementary Classroom

About the Project

Station-based instruction is a great teaching strategy for technology. Students move among a variety of activities matched to their ability or interest levels. Meanwhile, the teacher is freed up to work with small groups or individual students on special challenges.

Lesson Overview

LESSON RESOURCES

Activity Card Template: chrm.tech/sec3-045

Watch a 1st grader scan: chrm.tech/sec3-046

QR Scanner App for Chromebooks: chrm.tech/sec3-047

Blog post: Setting up digital workstations: chrm.tech/sec3-049

Set up three to five station activities for individual students or small groups to work on such as spelling games, flashcards, typing activities, or manipulatives. Students rotate through the stations every 5 to 15 minutes. Lessons usually take place over at least a 30-minute period. Ideally, one of the stations involves working directly with the teacher.

Tips and Suggestions

Getting students to the right website, screen, page, etc. can be a slow process. Speed it up by using QR codes and access cards so that elementary students are taken to the correct page each time. Develop sets of activity cards for your workstations, so that you can quickly pull them out as needed. Students can use a QR scanner app and the camera on their Chromebook to visit the activity website.

Remember, not all of your stations need to incorporate technology! Having a variety of digital and hands-on activities will increase student interest.

These Are A Few Of My Favorite Things

About the Project:

Students get to introduce themselves to their classmates through a digital presentation. This activity works well during the first few weeks of school, as students are getting to know one another. It teaches students basic technology skills, multimedia manipulation, and can be used to practice biographical writing. It also lets students experience the power of collaboration and communication within a busy whole-class presentation.

Lesson Overview:

Begin by asking students to write a short autobiography. The length and complexity of this autobiography can be adjusted based on student ability. You may want to provide students with a list of things to include (favorite color, place, food, animal, family member, etc.). Distribute a shared Google Presentation to the class and assign one slide to each student. Each student is responsible for adding their autobiography to their assigned slide, along with multimedia images representing a few of their favorite things. After students have finished, they can look at the slides created by their classmates and add comments and notes of encouragement.

Tips and Suggestions

Break up this assignment into several steps: writing, collecting multimedia, designing slides, and viewing and commenting on classmates' work. Google Classroom is a great tool for spreading this assignment into multiple steps over several days and tracking student progress.

Use the research tool in Google Presentation to easily find and add multimedia.

LESSON RESOURCES

Class Presentation Example:
chrm.tech/sec3-050

Google Presentation Template:
chrm.tech/sec3-051

Google Research Tool:
chrm.tech/sec3-052

Elementary e-Books

About the Project

Students take great pride in their finished, published work and enjoy sharing their creations with others. Build on this natural motivation by helping students build easily sharable e-books. Not only will they get to show off their writing, they can add illustrations and animations to make their finished writing more attractive, dynamic and exciting!

Lesson Overview

LESSON RESOURCES
Student Example:
chrm.tech/sec3-053

Free Google Slide Templates:
chrm.tech/sec3-054

Custom page sizes in Google Slides:
chrm.tech/sec3-055

Select one of your students' writing projects. Fiction and autobiographical assignments work well, but any genre could be used. Have students create a new Google Presentation and treat each slide as a page in their book.

Students can insert images to illustrate their story or use a drawing program to create their own. You may also scan or take pictures of student illustrations to use. Students may layer and animate their illustrations as desired.

Coordinate a literacy night or book party for students to share their finished work with their classmates and family members.

Tips and Suggestions

Separate the writing process and the design process. Don't begin the book project until students have completely finished their written work.

Spend time showing students the template options available in Google Slides. This is also a good time to talk to them about multimedia and finding images for their books. At this stage, it's also important to think about the final format of the book. Will you print them out? Read them digitally?

Finally, make sure to build in time for students to share their work with one another.

Lesson submitted by Debra Nichols, Music & Technology Teacher Milan Area Schools, Michigan

Colors in Any Language

About the Project

This lesson is designed to help young students learn about different languages and cultures. It features a blending of language instruction, social studies and art. This project can be modified into a small group project for two to three students.

Lesson Overview

LESSON RESOURCES
Student Example:
chrm.tech/sec3-056

Blank Template:
chrm.tech/sec3-057

Colors in other Languages:
chrm.tech/sec3-058

Google Research Tool:
chrm.tech/sec3-059

Create a Google Presentation with one slide for 5 to 10 colors (red, green, blue, orange, etc.). Use Google Classroom to distribute the template file so each student has their own copy. Ask students to find the translation for each color in the language you are studying (French, Spanish, etc.) and add it next to the English word. They can also be asked to find or write a sentence in the language on each slide and include a related image or video.

Depending on time constraints, add mini lessons on research, copyright and citations, multimedia, and slide design elements.

Tips and Suggestions

This works well as a collaborative activity where all students are working together on the same presentation. Set aside time for students to look at the slides created by their classmates. You can allow them to leave comments. And provide coaching on helpful and appropriate feedback.

Lesson submitted by Mary Hankins, K-5 Technology Teacher, Owosso Public Schools, Michigan.

Dream Bedroom

About the Project

Students get to research, plan, and design their own "Dream Bedroom" while learning basic budgeting and spreadsheet skills. This lesson will help them build smart consumer skills as well as learn how to budget money wisely.

Lesson Overview

Begin by engaging students in a whole-class discussion on bedroom "needs" vs "wants". Agree as a class on the necessary purchases that must be made before "wants" can be considered. Students then visit websites to collect images of items for a dream bedroom space. GIve them a budget and have them track how much each item costs using a Google Drive spreadsheet. Show them how to search online for the best price on bedroom items. Finally, help students make decisions about what they can afford. An optional extension is to have students collect the images of their Dream Bedroom items and showcase them in a multimedia project. Google Presentation, Google Drawing, LucidPress, and Canva are all good options for creating visual displays.

LESSON RESOURCES

Lesson Template:
chrm.tech/sec3-060

Finished Example:
chrm.tech/sec3-061

The Google Drive spreadsheet can be developed by the teacher and distributed via Google Classroom, or students can be asked to build it on their own. Advanced students can be challenged to use simple spreadsheet functions (=sum) to automatically calculate prices in each column.

Tips and Suggestions

• Google spreadsheet configured with necessary columns

• A list of bedroom "needs" (can develop this list as a class)

• A list of home furnishing websites for students to use (can develop as a class)

Lesson submitted by Mary Hankins, K-5 Technology Teacher, Owosso Public Schools, Michigan.

City Builder

About the Project

This lesson teaches students the basics of data collection, estimation, and visual design by asking them to plan and design a city using Google Forms and Drawing. This project can easily be expanded to emphasize math, technology, civics, and art skills.

LESSON RESOURCES
Student Instruction Template:
chrm.tech/sec3-062

City Creator:
citycreator.com (free online city builder tool)

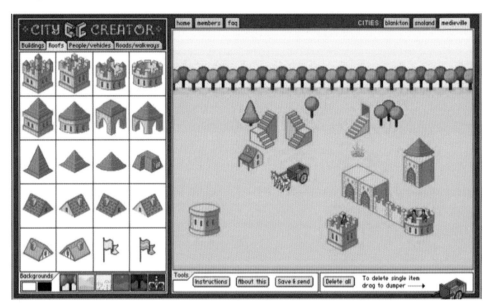

Lesson Overview

Students start by identifying all of the major features a city is responsible for managing or providing, such as streets, parks, schools, police and fire stations, hospitals, water sources, etc. Using Google Docs, students then build a city budget. A set dollar amount can be given or students can be given the freedom to spend as much as they choose. They list how much money will be spent on various items needed for their city. Once students have finished their city budgets, they get to build their city. This can be done using Google Drawing, City Creator, or another design program.

Tips and Suggestions

It can be helpful to have students plan their city on paper before building it digitally.

Lesson submitted by Ashlea Weston, Bethlehem Lutheran School, Indiana

Character Analysis with Venn Diagrams

About the Project

The Venn Diagram is a classic visual organizer for comparing and contrasting two or more things. Google Drawing is a marvelous tool for developing and completing Venn Diagrams. While this lesson is focused on language arts applications, it can easily be adapted to any subject area.

Lesson Overview

LESSON RESOURCES
Venn Diagram
Template:
chrm.tech/sec3-064

Finished Student
Example:
http://chrm.tech/9

Have your class make a list of all of the characters from a book they recently read. This activity is great for a simple book review! Ask students which characters would be fun to compare with each other. Characters can be matched based on having strong similarities or strong differences.

Once your pairs have been identified, assign groups of two to three students to perform character analysis of a pair using a Venn Diagram. If this is the first time your class has ever used a Venn Diagram, you will need to spend some time explaining what it is and how to use it.

Have students list the distinguishing characteristics about their matched characters. Using this information, have them create a Venn Diagram in Google Drawing (younger students may need a template). What do their characters have in common? What traits are unique to each one? Were there any surprises?

Tips and Suggestions

If you have access to a printer, print out the finished diagrams and hang them in your classroom.

Lesson submitted by Debra Nichols, Music & Technology Teacher, Milan Area Schools, Michigan

Fast Food - Healthy or Not?

About the Project

Help students learn how to make healthy meal choices by researching the nutritional information for common fast-food meals. Students collect information in a spreadsheet, then create a chart comparing fat, calories, sodium, etc. This lesson teaches important spreadsheet and data analysis skills and can be integrated into a math or technology course.

Lesson Overview

Students create a spreadsheet to record the nutritional information of a basic meal from three fast food restaurants (Wendy's, McDonald's and Burger King, etc). After the data has been collected, they compare the nutritional content from the three menus and make judgments on which one is the most healthy and what options are best.

If time allows, students can create a graph or chart comparing fat, sodium, calories, etc. Teach students how to create a chart, including important elements such as a key, label, grid lines, etc. Extend the activity by asking students to research a meal made up of their favorite foods.

Tips and Suggestions

Create a template spreadsheet for students that includes columns for all of the necessary information they need to collect. Use Google Classroom to distribute a copy of the template for each student.

Lesson submitted by Mary Hankins, K-5 Technology Teacher, Owosso Public Schools, Michigan.

LESSON RESOURCES

Spreadsheet Template:
chrm.tech/sec3-066

Example of a finished assignment:
chrm.tech/sec3-067

McDonald's nutritional information:
chrm.tech/sec3-068

Wendy's nutritional information:
chrm.tech/sec3-069

Burger King nutritional information:
chrm.tech/sec3-070

Simple Elementary Journal or Portfolio

About the Project

Student learning and growth is often slow and hard to observe, especially for the students themselves! Journals and portfolios are great ways to demonstrate their development throughout the school year. Both of these tools have a reputation of being a lot of work to set up and maintain, but Google Presentations provide a simple, effective, and fun way to track student progress, Best of all, they are easy to share with families and classmates!

Lesson Overview

A Google Presentation is the core of this assignment. You can have students start from scratch with a blank presentation, or, if you know some of the activities you will work on throughout the year, you can set up a template with major assignments and writing prompts already in place.

Google Classroom is a great tool for setting up and distributing portfolio or journal templates to your students. Once created, the template will be available in each student's Google Drive account, and they can reopen and continue working on it throughout the school year.

Students can take advantage of the webcam on their Chromebook to take pictures and insert them into their presentation. This also works well for capturing assignments that are done on paper (just hold it up to the camera and snap a photo!). If you do a project using another web-based tool, you can take a screenshot of it, or insert a link to the finished project.

LESSON RESOURCES

2nd Grade Journal Example:
chrm.tech/sec3-071

Elementary Journal Template:
chrm.tech/sec3-072

Professional portfolio example:
chrm.tech/sec3-073

Video Tutorial: Creating a presentation portfolio:
chrm.tech/sec3-074

Creating portfolios with Google Apps:
chrm.tech/sec3-075

Tips and Suggestions

Spend time explaining that this project will take the entire year to complete and that students will be adding to it regularly. Allow them to customize the theme and styling of their presentation to fit their personalities. You can begin by having them create a title slide with their name, photo, etc. Make sure that students are familiar with the basics of adding new slides, inserting pictures, etc.

Plotting Earthquake Data Using Google Draw

About the project

This inquiry-based activity introduces students to concepts of plate tectonics and boundaries. By plotting earthquake data on Google Draw, students discover the location of plate boundaries. They will also gain experience reading and working with complex sets of data.

LESSON RESOURCES

Earthquake Drawing Template:
chrm.tech/sec3-076

Sample Finished Drawing:
chrm.tech/sec3-077

Earthquake data from USGS:
chrm.tech/sec3-078

Modified sample data set from USGS:
chrm.tech/sec3-079

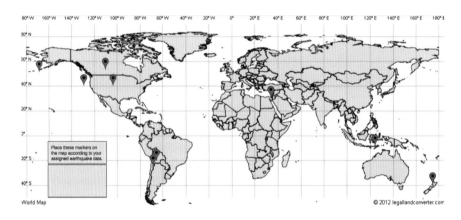

Lesson Overview

Create a Google Drawing that contains a world map with lines of latitude and longitude. The more lines of latitude and longitude, the better.

Go to the USGS earthquake data website and select earthquake data over a period of time. You may want to limit the magnitude range of earthquakes from 3.0-10.0. Download the CSV file for the earthquake data and print it out. Cut data into a manageable amount of events for each student (10 earthquakes per student works well) and distribute the sheets. Have students find the longitude and latitude for each event, then plot that earthquake data on the map in Google Draw.

An optional extension would be to have students share or print their finished maps and compare them with those of their classmates, then try to identify the fault lines where earthquakes occur most frequently.

Tips and Suggestions

Students will need to be familiar with latitude and longitude lines and how to plot a point when given this information. Providing students with a printed copy of their earthquake data keeps them from having to switch between their map program and the data set. Google Classroom is helpful in distributing the drawing template to students.

Lesson submitted by Meredith Nickerson, Dexter Schools, Michigan and Justin Ellsworth, Farmington Schools, Michigan

Numbers & Spelling with Sheets

About the Project

Google Sheets is a surprisingly effective way to teach elementary students basic spelling and number patterns. The clear division of the sheet into cells and the ability to move around the page using the arrow keys make these lessons easy for PK-K students to learn math, spelling, and basic technology skills.

Lesson Overview

Create a Google Sheet with content that you want your young students to work with. You might ask them to count by 1s, 2s, 3s, etc by filling in the missing numbers. For spelling, you can ask students to type out spelling words to practice letter recognition and keyboard literacy. You can also ask them to highlight the correct word, change the text color, etc. These activities can be easily adjusted for complexity over time and make great station activities.

Tips and Suggestions

Delete unneeded cells to minimize the potential for students to get "lost" in the spreadsheet. For spelling practice, you can use conditional formatting to automatically turn a cell green for a correct answer or red for a wrong answer.

Use Google Classroom to distribute a copy of your Sheet template to all of your students. You can also let students work collaboratively. If using this lesson in a station setup, use revision history to quickly reset your Google Sheet back to its original state.

Lesson submitted by Kunio Ouellette, Essexville, Michigan.

LESSON RESOURCES
Math Number Assignment Template: chrm.tech/sec3-080

Spelling Assignment Template: chrm.tech/sec3-081

Conditional Formatting in Google Sheets: chrm.tech/sec3-082

Lesson Ideas for Middle and High School Students

Students in middle and high school will use Chromebooks to learn and practice skills in various content areas. While many of the lessons in this section are presented with a focus on a specific subject area (math, science, language arts, etc), with a little creativity, many of them can be modified to fit a wide range of content areas.

When reading through this section, try not to focus too much on the specific details of the lesson, instead use these sample lessons to spark your imagination for how you can incorporate Chromebooks into your classroom using your unique skills and instructional style.

Create Your Own Home Page

About the Project

Students love to customize their Chromebooks. Let them! During this lesson, students will create a personal home page with access to all of the tools and information important to them. This lesson is ideal for the very beginning of the school year, when students are still learning about their Chromebook and how to work with it. Use this lesson as an opportunity to build important skills that will be used later in the year.

Lesson Overview

Develop a list of required elements for each student's home page. For instance, you may require a link to Google Classroom, your gradebook login, media center databases, or any other resources you plan to access on a regular basis.

Decide if you want all students to use the same tools to create their home page, or if you want to give them the ability to pick from a list of options. Here are some good tools to consider:

Google Sites - This is a great opportunity to introduce students to the basics of creating and editing a page, especially if you plan to use Google Sites for assignments later in the year.

Symabloo - A super-easy tool for creating a list of frequently used sites. Great for elementary and middle school students. Get started at symbaloo.com.

Start.me - A true custom home page tool with widgets, apps, bookmarks, and more. Great for high school students who want to stay connected with many sources of information. Try it out at start.me.

Tips and Suggestions

Students can easily spend hours customizing their sites. Give them a specific list of elements to include and a firm deadline for completing the home page assignment. Let them know that they can always tweak it and change things up throughout the year, but they need to finish a basic, functional model on time to get a good grade.

Once completed, students should bookmark their page for easy access and/or configure their Chromebook to automatically open it as their starting page. To do so, visit Settings > On Startup. Note that the "on startup" feature may be locked by your IT department, depending on your device configuration.

LESSON RESOURCES
Sample Google Site Start Page: chrm.tech/sec3-154

Sample Symbaloo Start Page: chrm.tech/sec3-155

Sample Start.me page: chrm.tech/sec3-156

Student Assignment Template: chrm.tech/sec3-157

Get Cultured!

About the Project

Pairs of students work on a cross-curricular project which blends culture and language. Not only does this assignment build skills related to language usage and composition, it introduces students to distance collaboration.

Lesson Overview

LESSON RESOURCES
Easybib add-on for Google Docs:
chrm.tech/sec3-084

Easy Accents add-on for Google Docs:
chrm.tech/sec3-085

Orange Slice Rubric Creator:
chrm.tech/sec3-086

Pair up students from language arts and world language classes. Each pair is responsible for researching a specific person, place, country, or topic (cultural icon, form of government, literature, current event, historical topic, etc.), based on your curriculum.

Students share their ideas, work, research, and final product using Google Drive, including a written report on Google Docs. They share their work with the class through a two-minute talk using Google Presentation.

Tips and Suggestions

Walk students through the process of creating a folder in Google Drive and sharing it with their partner. Explain that part of this assignment's purpose is to empower them to collaborate effectively from a distance.

Give older students the freedom to move beyond Google Presentation and try another tool, such as Prezi, Pear Deck, Nearpod, PowToon, etc.

Lesson submitted by Valerie Johnson and Kimberly Jasper, Milan, Michigan

Create a Collaborative Study Guide

About the Project

The best way to learn something is to teach it to someone else.

For this activity, students who need to learn materials for an upcoming test will use the Ziteboard whiteboard extension to solve a math problem while recording their solution using the extension Screencastify. Students then post their recording into a collaborative Google Slides presentation to share with the class.

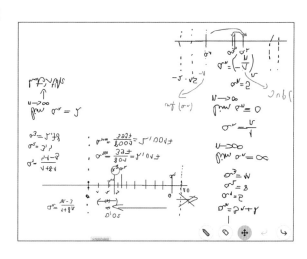

LESSON RESOURCES

Ziteboard Chrome extension:
chrm.tech/sec3-087

Screencastify Chrome extension:
chrm.tech/sec3-088

Math Student Tutorial Examples:
chrm.tech/sec3-089

Lesson Overview

Give each student a different problem to solve and time, if needed, to explore Ziteboard and Screencastify.

Once students have finished, have them find the Screencastify folder in Google Drive and locate their recording. Students will need to change the share settings on the recording to "share with anyone with the link."

The final step is to have students place their links into a collaborative class Google Slides presentation. This can also be done in Google Sheets or Docs; however, Google Slides will allow you to assign each student their own Slide page.

Each student now has access to a student-generated guide!

Lesson submitted by: Carrie Moeggenberg, Instructional Technology Coach, Ludington, Michigan.

Flipped Book Reviews

About the Project

Have you ever sat through a series of book reports or student presentations and watched the clock tick away as one student after another comes to the front of the room to give their report? Have you ever longed for a more efficient way to allow students to present their work? With flipped book reports, students simply record and post their presentations online, which can then be easily shared with family members and classmates.

Lesson Overview

For this activity, students will develop a book report based on the guidelines provided by the teacher and use the free Chrome app Screencastify to record their voice and screen as they deliver their presentation. It may be best to assign the recording as a homework assignment to minimize background noise and distractions.

Students make their recordings available to the rest of the class. Posting videos to Google Classroom is a simple way to exchange video links! Then, each student watches and peer edits several videos created by classmates. A rubric for evaluating should be used to help guide the peer editing process.

Lesson submitted by Jim Kelly, Director of Curriculum and Technology, Stockbridge, Michigan

LIMITED RECORDING
Note: The free version of Screencastify limits recordings to a maximum 10 minutes.

LESSON RESOURCES
Introduction to the Flipped Classroom:
chrm.tech/sec3-094

Screencastify:
chrm.tech/sec3-095

Sending a video to Google Classroom:
chrm.tech/sec3-096

Orange Slice Rubric Creator:
chrm.tech/sec3-097

Simple Story Builder

About the Project

Dialog and character interaction is what makes fiction immersive and engaging. Learning how to write engaging dialogue is difficult, but with Google Story Builder, students can practice the writing process in a fun and interactive way. Students can add multiple characters and background music to set the tone for their story.

Lesson Overview

LESSON RESOURCES

Google Story Builder: chrm.tech/sec3-098

Sample story which illustrates the importance of punctuation: chrm.tech/sec3-099

Random Plot Generator: chrm. tech/sec3-100

Using the Story Generator in the Classroom: chrm.tech/sec3-101

Story Planner Template: chrm.tech/ sec3-102 .

The purpose of this assignment is to help students practice and experience writing dialogue. It takes a lot of time to come up with characters, a plot, and a setting for a short story. Rather than coming up with these elements on their own, have students use a plot generator to automatically suggest all of the required elements. Not only will this save time, students will find the crazy plot suggestions quite entertaining!

After generating and reviewing their plot, students will use the Google Story Builder tool to write the dialog for a portion of their story. The story builder tool is purposefully restrictive and only allows a maximum of 10 characters, with no more than 10 comments from each one.

After completing the dialog, students can customize their background music, then publish and share the video for others to watch.

Tips and Suggestions

Stories cannot be modified once they are published. Encourage students to plan their story on paper before building and publishing their digital story.

Stories cannot be downloaded. The only way to share a story is to copy the unique link to a published video. Create a shared class document to collect and share links so that students can watch each other's creations.

Lesson submitted by Carrie Moeggenberg, Instructional Technology Coach, Ludington, Michigan

Design a Restaurant

About The Project

The primary purpose of this lesson is to allow students to research, imagine, and create a design of their own. Students will get to practice research and organization skills, planning and project management concepts, as well as creativity and problem solving. Creation is a powerful learning experience, so while this particular lesson focuses on designing a restaurant, it can be modified to fit any other objective.

From a curricular perspective, this lesson can be adjusted to focus on math concepts (designing to scale, measurement conversions, volume and surface area), language arts (writing descriptions, marketing material, etc), social studies and culture (looking at different layouts based on country or cultural differences), and technology skills (using spreadsheets, prototyping, web-based tools, etc).

VR Addition

For an added splash of excitement, consider using this lesson to introduce your students to virtual reality! Google Cardboard is an inexpensive virtual reality viewer that is compatible with nearly all mobile phones. Download the app YouVisitVR (available for iOS and Android) to find immersive 3D video of restaurants from around the world. Students will feel as if they are actually there!

You don't need Google Cardboard for every student. Just a few will provide an opportunity for students to experience VR for the first time.

Lesson Overview

This lesson is divided into three parts: research, plan, create.

Allow students to investigate various types of restaurant floor plans. This can be done through Google Image search, watching videos, or visiting yelp.com. Provide students with a guided worksheet to help them make observations about different types of restaurants. Students should insert their sketches into a Google Doc and select one design.

LESSON RESOURCES
Restauraunt
Observation
Worksheet:
 chrm.tech/sec3-103

Sketchpad Chrome
App:
chrm.tech/sec3-104

Floorplanner Chrome
App:
chrm.tech/sec3-105

Restauraunts
available from
YouVisitVR:
chrm.tech/sec3-106

Blogpost: Getting
Started with VR:
chrm.tech/sec3-107

Learn more about
Google Cardboard:
chrm.tech/sec3-108

Based on their observations and research, students will design their own restaurant. Begin by using a free drawing tool like Sketchpad or Google Drawing to design a preliminary design. The final design can be created using Floor-planner, a Chrome app that provides all the tools you need to create attractive floor plans including walls, doors, and furniture.

Save the floor plan and insert it on the Google Doc with the rough sketches, or use a screen capture tool (Screencastify or the integrated Chrome OS capture function) to grab an image of the completed floor plan.

Finally, students should create a restaurant overview document that includes the name of their restaurant, the floor plan, and a brief description of the dining experience they have created.

Lesson submitted by Tonya Nugent, middle school teacher, Commerce, Michigan

Close Reading of Informational Text

About The Project

An area of emphasis in the Common Core state standards is the identification of an author's point of view and identification of textual evidence which supports or rejects their claim. Students must able to closely read and analyze information for key ideas, biases, and supporting evidence. This assignment provides students an opportunity to practice close reading skills while exploring a controversial topic.

The principles and objectives on which this lesson is based can be easily adapted to any controversial topic, such as genetically modified foods, raising the minimum wage, or naturalization.

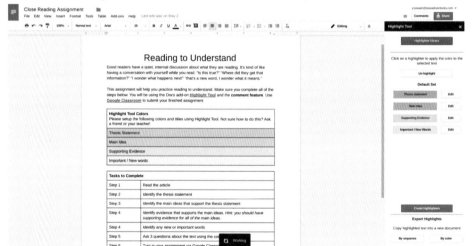

LESSON RESOURCES

A Non-Freaked-Out, Focused Approach to Close Reading:
chrm.tech/sec3-109

Close Reading Assignment template:
chrm.tech/sec3-110

Completed close-reading assignment:
chrm.tech/sec3-111

Sample comparative presentation:
chrm.tech/sec3-112

Lesson Overview

Provide students with three to five sources of information. While the focus of this lesson is on written information, you can mix in video and audio sources as well. This assignment can be completed individually or as a small group project (2-3 students per group).

NewsELA and **DOGONews** are good sources to find articles organized by age and reading level. Provide students with copies of the informational text. **Google Classroom** is an easy way to distribute this information.

Students will use one or more of these annotation tools to mark up the original document. It may be helpful to provide students with a list of items to look for as they read.

Highlight Tool: This simple add-on for Google Docs provides students a simple way to highlight text using multiple colors. They can also label and collect their highlights into a new document. Install highlight tool for Docs: chrm.tech/sec3-170

Kami: Kami allows students to highlight, annotate, and draw on top of information in PDF format. Kami is collaborative, allowing multiple individuals to mark up the same document. Install the Kami Chrome App: chrm.tech/sec3-171

Commenting in Google Docs: The integrated comment tool in Docs provides an easy way for students to share their own observations about the text. Learn more about commenting in Docs: chrm.tech/sec3-172

Tips and Suggestions

Students benefit from having guiding questions to consider as they read through the source material. A worksheet can help them remain focused on the central theme of the lesson.

An optional extension to this activity would be to have students create a presentation or summary comparing the themes, evidence, and biases of the sources that were read. Presentations can be created using **Google Presentation**, **Prezi** or any other presentation tool.

Lesson submitted by Amy J. Weston Middle School Language Arts Teacher, Walled Lake, Michigan

History of ANYTHING Interactive Museum

About the Project

History can be a dry subject, especially when taught by lecture. What if students learned and then taught each other the material through interactive museum exhibits, then assessed each other's learning? What if they were empowered to teach other classes and even adults - teachers, family members, and community members?

This project lets students build their historical knowledge, research skills, collaboration skills, and teaching skills by curating their own virtual museum! It can be used at any grade level with as much or as little direction on topics or resources as needed. Besides being a terrific learning project, it gives students the chance to engage in the real-life work of historians, museum exhibit creators, and docents.

Lesson Overview

Students work in groups of two to three in defined roles to create and then present an exhibit on a person, place, or thing from your unit of study. Each group can be assigned to a chapter, section, person, place or event from the time period being studied. You will serve as facilitator of the project and have 'curator' meetings to check on how groups are doing and provide support. It's helpful to provide progress targets along the way. Don't forget to celebrate the final products on exhibit day!

Tips and Suggestions

Before beginning this assignment, clearly explain the required elements of the project (pictures, video, vocabulary words, etc.). It is also important for your students to know and follow the rules of conducting research, using proper citations, and respecting image copyrights.

Students will need three tools to complete this project: a collaboration space, image tools, and an exhibit space. The options listed below are a good place to begin, however, many other tools are available.

COLLABORATION SPACE

Students will use this space to collect and share information and to communicate with members of their group. Possibilities include:

LESSON RESOURCES
Lesson Template & Rubric:
chrm.tech/sec3-116

Video of a Cold War "Museum":
chrm.tech/sec3-117

Video of a 1970s "Museum" exhibit:
chrm.tech/sec3-118

Augmented Reality (AR) in the classroom:
chrm.tech/sec3-119

Integrating AR into your classroom:
chrm.tech/sec3-120

Interactive Exhibits with QR codes:
chrm.tech/sec3-121

Google Drive (great for storage, poor for communication)

Google Keep (great for short notes, to-do lists, reflections, and resource collection)

Trello (project management tool for dividing and tracking work in a group)

IMAGE TOOLS

Students will need to gather images from the web, create their own works, and photograph objects for inclusion in their exhibit. Possible tools include:

Screencastify (takes screen images and video)

Webcam Toy (use a Chromebook webcam to take a picture)

Pixlr, PicMonkey, SumoPaint (image editing tools)

EXHIBIT SPACE

Each student or group will need to create a visual display of the information collected. Decide if you want students to create digital displays (website, presentation, etc.) or print displays (brochure, poster, book, etc.).

Digital / Print Poster Tools: **Takk**, **Canva**, **LucidPress**, Glogster, S'more, Blend-Space

Video Tools: **WeVideo**, **PowToon**, **StoryBird**, **Movenote**

Presentation Tools: **Google Presentation**, Prezi, Pear Deck, Nearpod

Webtools: **Versal**, **Weebly**, **Wix,** Google Sites

Lesson submitted by Matt McCullough, HS History Teacher, Kalamazoo, Michigan

CHROME WEB STORE
Each of the products in bold in this section can be found in the Chrome Web Store. Just search for them by name!

Augmented Reality

Many assignments present an opportunity to mix digital and paper based content using QR codes and augmented reality (AR). Both technologies allow the linking of digital content (videos, images, documents, etc.) with physical objects.

Display a QR code with a real-world object to connect viewers to related videos and other digital content. QRstuff.com is a great place to create QR codes. The **Scan QR** Chrome app is great for reading QR codes. You can find QR code readers for any mobile device - just search for them in the iOS App Store or Google Play.

Augmented reality overlays digital content directly on top of physical objects viewed through a handheld device's screen. **Aurasma** is the leading tool for creating such AR experiences. Scan this QR code to see a video featuring **Daqri** augmented reality element blocks!

Cone Snail Classification (Taxonomy)

LESSON RESOURCES
Sample shell classification lab worksheet and instructions: chrm.tech/sec3-122

Shell pictures (single page, Google Drawing): chrm.tech/sec3-123

Shell pictures (slideshow): chrm.tech/sec3-124

Learn more about the Cone Snail chrm.tech/sec3-125

Learn more about Doctopus and Google Classroom: chrm.tech/sec3-126

Classification Infographic created in Piktochart: chrm.tech/sec3-127

About the Project

Science activities are perfect for utilizing the collaborative power of Chromebooks and Google Drive. In this activity, students will create a classification system for snail shells. Students will review the systems developed by the other lab groups in their class, allowing them to learn from the different perspectives of others. Take advantage of the collaborative power of Google Drive to pool lab data and provide students an opportunity to ask questions of one another and the lab process.

Lesson Overview

Provide students with a collection of cone snail shells in either physical or digital form. If you use a digital image of the shells, you may find it helpful to print it out for easier reference.

Explain that even professional scientists are still debating how to classify the more than 800 species of cone snails! This project gives students the chance to figure out their own approach, while learning about the dangerous venom the snails use to capture prey, and the exciting medical advances being explored.

Ask students to classify the cone snail shells based on what they can observe. Each system should have at least seven different categories, and none of the sample shells could fall under more than one. Criteria could include elements such as colors, whether the shells have spots or more than one row of markings, the direction of their swirl, their length-to-width ratio, etc.

After developing their own classification system, allow students to review the systems created by other groups. Then, share with them the systems currently being used by professional biologists.

INFOGRAPHIC BEAUTY
Build beautiful infographics using Piktochart! Available in the Chrome Web Store: chrm.tech/ sec3-129

Ask students to make observations about the different methods and difficulty of classifying these shells. The lesson concludes by asking students to build a chart representing the biodiversity of the shell collection.

Tips and Suggestions

DOCTOPUS
Doctopus is a great tool for facilitating group projects. Learn more: chrm.tech/ sec3-128

This assignment works best in small groups of two to three students. Doctopus, a free add-on for Google Sheets, will help you create and distribute all of the necessary resources to your lab groups. Doctopus integrates with Google Classroom, allowing you to import your class roster.

Invite students to use **Google Drawing** or **Sketchpad** to annotate on top of the snail shell image in order to identify family characteristics.

A possible extension for this lesson is to ask student to develop an infographic depicting the biodiversity of their shell families using Piktochart, a free infographic builder.

Lesson submitted by Shawn Maison, High School Science Teacher, Bay City, Michigan

The Science of Injuries

About the Project

Nothing is more simultaneously gross and fascinating as injuries to the human body. Students enjoy learning about the human body, especially when it comes to breaking it and putting it back together!

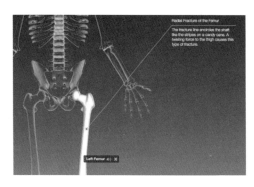

For this activity, you will use BioDigital Human, a free Chrome app that allows you to explore the human body in 3D. If you have ever used Google Earth, you will feel right at home with BioDigital Human.

Lesson Overview

Assign students to study an injury or medical condition (i.e. broken femur, collapsed lung). You might also allow students to research an injury they or a family member have experienced.

Ask students to use the controls to focus on areas of the body and discuss major elements of the injury. The can use the annotation tool to add text to point out important features.

Students will use the camera tool in BioDigital Human to grab an image of the completed body view. Insert the image into a Google Document (see template below) and have students include a narrative of the injury, treatment, and recovery prognosis.

Tips and Suggestions

BioDigital Human is pretty awesome, so make sure you build in some time for students to explore. The "conditions" tab in BioDigital Human will provide information on quite a few conditions and injuries. Make sure students know how to use the "edit" toolbar in BioDigital human to change the view, annotate, add notes, and take a picture.

This assignment could easily be turned into a screencast lesson, where students record a narrated tour of the body and their injury using Screencastify or another screen recording tool.

LESSON RESOURCES
Sample student worksheet:
chrm.tech/sec3-130

"Conditions" in BioDigital Human:
chrm.tech/sec3-131

The 5 Most Common Sports Injuries:
chrm.tech/sec3-132

Sports Injuries A-Z:
chrm.tech/sec3-133

Build Your Own Tip Calculator

About the Project

Computer programming, or coding, is a STEM skill that can be introduced at any age. Coding incorporates elements of math, logic, and design. In this activity, students will learn basic coding skills by creating a simple tip calculator "program" using Google Sheets.

fx	=SUM(C2*C3)		
	A	**B**	**C**
1	**Enter Your Information Below**		
2		**Bill**	$25.12
3		**Tip Rate**	0.25
4			
5	**Results**		
6		**Tip Amount**	$6.28
7		**Total Bill**	$31.40

Lesson Overview

LESSON RESOURCES
Tip Calculator Lesson Overview:
chrm.tech/32

Sample Calculator:
chrm.tech/31

Data Validation in Google Sheets:
chrm.tech/33

This activity works best when presented as an inquiry-based assignment in which students are given a problem to solve and some suggested tools and resources. The process and collaboration required to accomplish the task is just as important as the final result.

Challenge students to figure out how to create a tip calculator that allows users to enter values into blank cells and have the Sheet output the tip amount and total amount to pay the restaurant.

Tips and Suggestions

The tip calculator project is appropriate for most middle school students. Challenge more advanced students to build a "what if" grade analyzer that calculates their overall class grade "if" they get XX on their next assignment.

Lesson submitted by Kevin Brookhouser, Code & Design Teacher, Monterey, CA

Mapping Concepts Creatively

About the Project

The more students engage with complex topics, the more challenging it becomes to understand and visualize the relationships between concepts. Concept mapping provides an opportunity to physically arrange ideas in logical order.

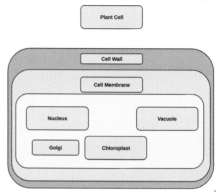

Besides helping students identify the relationships between ideas, concept mapping provides valuable insights into how students are thinking. That can help you identify major misconceptions before testing time.

LESSON RESOURCES

Google Drawing Template:
chrm.tech/sec3-134

Sample finished examples:
chrm.tech/sec3-135

3-Circle Venn Diagram Template:
chrm.tech/sec3-136

Timeline Template:
chrm.tech/sec3-137

Lesson Overview

Identify key terms, skills, names, places, dates, etc. that your students need to know. Select a concept mapping tool (see suggestions below) and place these items as individual shapes on the canvas. Instruct students to organize the objects. Do not give them any additional instruction, advice, or hints. Let them struggle through the cognitive process of creating order from the provided items. Depending on the topic, however, you may want to provide some initial structure to your concept map. For example, in a history class, you may want to include a rough timeline on which students can place important people, events, battles, etc. In a science classroom, you may offer a Venn diagram for organizing terms.

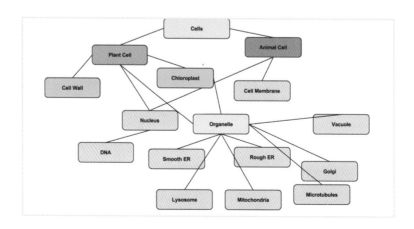

Tips & Suggestions

This assignment works best as a formative assessment lesson which can be adapted for all grade levels and subject area.

Be prepared to see some very creative organizational strategies! The beauty of this assignment is that it has no correct answer - and students have the freedom to come up with their own solutions, some of which you may not have considered. Grade students on the process, not their final solutions.

Concept Mapping Tools:

Google Drawing - Simple yet effective. Integrates with Google Classroom, which makes distributing templates very easy.

Collabrify Map - A product specifically developed for the K-12 classroom, featuring the elements of Google Drawing. Excellent for younger children.

MindMup - Simple to use; no signup or login required. Connects with Google Drive. Great for elementary and middle school students.

Lucid Chart, Mind Meister, Draw.io - Very powerful tools with complex shapes and relationships and a large number of templates. They may be overwhelming for younger students.

Each of the tools in bold offer a Chrome Web App. Search for them by name in the Chrome Webstore.

Virtual Field Trips

About the Project

Why talk about other countries when you can see them for yourself? Bring history and culture to life by allowing students to visit and explore natural wonders, historical monuments, museums and far away places right from your classroom.

LESSON RESOURCES

How to plan a virtual field trip:
chrm.tech/sec3-138

Sample Student Worksheet:
chrm.tech/sec3-139

Bulding your own Lit Trip:
chrm.tech/sec3-140

Student generated lit trips:
chrm.tech/sec3-141

Lesson Overview

Creating a virtual field trip takes planning and preparation, just like a real field trip! Give yourself plenty of time to gather resources and develop lessons for your trip. The more you plan, the better the experience! Each trip will incorporate the following elements:

TRAVEL ITINERARY

Students need to know where they are going, how they are getting there, and how they need to manage their time. Provide students with clear instructions on how they will take their tour (Google Earth, Maps, websites, etc.) and what "stops" they will make along their tour. Bonus for math students: budgeting for hotels, transportation, restaurants, gear, etc. while keeping track of local currencies.

SOUVENIRS AND ARTIFACTS

Set clear expectations on what "souvenirs and artifacts" your students will create or bring back after their "trip." This might include completing a worksheet, taking screenshots, or collecting information to incorporate into a presentation, video, or website.

Tips & Suggestions

A growing collection of tools and resources will help your students take a virtual trip. Here are some options to consider:

Google Lit Trips: A collection of free virtual field trips based on popular novels read in grades K-12, including *The Diary of Anne Frank*, *The Kite Runner*, and *Make Way for Ducklings*. They include everything you need to take a virtual field trip! Learn more: chrm.tech/sec3-143

Files downloaded from the Google Lit Trip Library and opened on a Chromebook will open in Google Tour Builder or My Maps.

My Maps: Create a custom tour using My Maps by Google. You can create custom place markers, add questions, videos, images, and more! My Maps is integrated with Google Drive, and finished maps are easy to share with students. Learn more: chrm.tech/sec3-144

Google Tour Builder: An easy-to-use tool to develop your own virtual tours. This tool would be easy for students to use to create and share their own virtual field trips. Tour Builder also features a gallery of tours created by others. Learn more: chrm.tech/sec3-145

Google Street View: Walk along the streets of Rome or visit Stonehenge in England. Street View features street-level photography of thousands of locations across the globe. You really feel like you are there! Learn more: chrm.tech/sec3-146

IN THE CLASSROOM
Watch how social studies teacher Bobby Lake uses Tour Builder to get his students excited about history: chrm.tech/47

GOOGLE EARTH SUPPORT
While Google Earth has been a long-time favorite of both teachers and students, it is not compatible with Chromebooks. Google Earth tours can be exported as .kmz files which can be viewed using Google Tour Builder. Learn more: chrm.tech/sec3-142

EarthCam: Check out the world in real-time using EarthCam, a massive network of live webcams from New York to New Zealand and beyond.

Virtual Scavenger Hunt: This tool is ideal for teaching history, geography, and science alike! Create a global scavenger hunt for important sites related to your subject, and let students collect "coins" for every one they find. chrm.tech/sec3-148

Google Cultural Institute: Interested in art, architecture, or historical artifacts? The Google Cultural Institute provides the ability to create your own custom museum exhibit of famous works of art and primary source material from around the world. You can also take a virtual tour of some of the world's most amous museums, like the National Portrait Gallery in Washington D.C. Learn more: chrm.tech/sec3-153

Virtual Reality in the Classroom

Virtual Field trips provide an opportunity to incorporate Virtual Reality (VR) into the classroom. Google Street View and and Google Expeditions provide great opportunities for students to immerse themselves in a new place and culture. VR experiences require access to a mobile device (iOS or Android) and a VR viewer such as Google Cardboard. Students can experience spaces with their mobile device and complete their "artifacts" using a Chromebook.

Virtual Reality Resources

Blogpost: Getting Started with VR: chrm.tech/sec3-149

Learn more about Google Cardboard: chrm.tech/sec3-150

Learn more about Google Expeditions: chrm.tech/sec3-151

Sign up for Google Expeditions: chrm.tech/sec3-152

Math & Science Infographics

About the Project

Math and science courses frequently require students to work with large amounts of data. While some students have no problem dealing with abstract numbers, some students struggle to visualize the true impact of the data on their screen. An infographic is a fun way to turn abstract data into a colorful, impactful display.

For decades, teachers have been asking students to create visual displays. From poster boards, to flip-books, to tri-fold brochures, these displays provided students an opportunity to organize complex information in an easily digestible format. An infographic is a modern adaptation of these paper-based assignments that accomplishes the same goal of teaching students to organize complex information without the necessity of rubber cement and glitter.

FREE APPS!
Piktochart and Infogram are both free Chrome apps specifically focused on creating beautiful and engaging infographics.

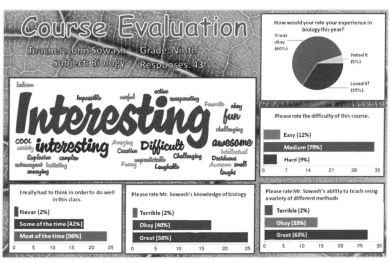

Lesson Overview

The key element to an infographic is the underlying data. This data can come from many sources: lab experiments, community surveys, or publically available data sets. The first step in this assignment is to collect and organize the data.

Students will also need to make determinations about the conclusion or key findings their data illustrates. This is a great opportunity to talk about statistical significance, data integrity, data-driven conclusions.

After students have collected and analyzed their data, give them examples of quality infographics to review and evaluate. How was the data displayed? What types of charts and graphs were used? How was the data titled and captioned?

Finally, ask students to create their own infographic. Their infographic should be clear, easy to read, and accurately reflect the data that was collected.

Tips and Suggestions

This assignment can be extended over the course of several days (collect, analyze, create) or condensed into a single class period. If time is limited, provide your students with a data set rather than asking them to collect their own data. Google Trends and Google Correlate provide lots of interesting data sets that can be analyzed (check out searches for "flu symptoms" by geography and time to see if you can pinpoint a flu outbreak in the United States).

There are endless tools available for students to use to assemble their infographic. Give your students clear time parameters for this assignment as it is easy to spend days editing and adjusting an infographic. Here are a few recommended tools:

Piktochart - A free Chrome apps specifically for creating infographics. Very nice visual displays with lots of free templates.

Infogram - A free Chrome app for developing interactive infographics that allow a viewer to click and explore live data. Requires students to load their data set into the infographic.

Google Drawing - The infographic above was created in Google Drawing. The graphs are screenshots from the automatically generated displays created by Google Forms.

Google Presentation - The new chart feature available in Google Presentation makes it easy to insert a live chart based off data from a Google Sheet.

LESSON RESOURCES
Google Correlate:
chrm.tech/sec3-164

Google Trends:
chrm.tech/sec3-165

Google Public Data Explorer:
chrm.tech/sec3-166

Information is beautiful:
chrm.tech/sec3-167

Nerdgraph:
chrm.tech/sec3-168

Infographic showcase:
chrm.tech/sec3-169

Conclusion

Welcome to the end! Having read the previous pages, I hope you are beginning to gain a vision for how technology can be used to enhance the classroom.

Remember: technology by itself does not redefine effective instructional practice. Instead, it opens up new opportunities to teach our students the skills of creativity, collaboration, and communication. Chromebooks are designed for all three. Chromebooks do not isolate students, they connect them and allow them to combine their creative efforts in new ways. This book, for instance, instead of being written alone in a dark room, came to light through the combined efforts of more than 60 educators from around the country. That would have been inconceivable without the collaborative technologies of Gmail, Google Drive, and various Chrome Apps such as Kami and Pixlr, all accessed using a Chromebook Pixel.

You may still be wary. After all, for decades, educational technology has been complicated, cumbersome, and challenging to effectively integrate. That is starting to change. It isn't necessary to commit to "go paperless" or "flip your classroom" in order to get started with Chromebooks. Start small. Look for easy wins and simple opportunities to expand your instructional strategies and learning opportunities through the use of tech.

Chromebooks represent classroom technology that can be invisible, blending into the learning environment. Removing complexity in favor of exploration and creativity, Chromebooks represent an ideal classroom device for students, teachers, and IT administrators.

Many educators also fear that classroom technology diminishes the role of teachers. In fact, the opposite has happened. With such a large array of tools readily available for use in the classroom, we need highly trained, highly skilled educational professionals to make the careful decisions necessary to improve student learning. Oh, and don't be afraid to put the technology away from time to time and have a good ol' fashioned class discussion. Or write something with a pencil on paper (gasp!).

Now it's your turn. Take what you've learned from this book and use it to transform your classroom. I would love to hear about it! Share on social media using #chromebookEDU or drop me an email at jrsowash@sowashventures.com. Chromebooks are an expression of all that is great in education - creativity, organic exploration, and a thirst for knowledge. Harness their potential as you guide your students to do amazing things.

- John Sowash

June 2016

Ready for more?

Stay connected and up to date on the latest Chromebook updates and opportunities!

Stay Social

Twitter: Follow and post using #chromebookEDU

Google+: Follow the Chromebook Classroom Collection curated by John Sowash. This collection of posts will help you stay up to date on the latest developments to the ChromeOS platform: - chrm.tech/7q.

Join the Chromebook EDU community with over 20,000 members. This is a great forum to ask questions and discuss the latest applications for Chromebooks: chrm.tech/7r

Professional Development

ON-SITE PROFESSIONAL DEVELOPMENT

John Sowash and the gEducator team of trainers works with schools around the country, helping them use technology in meaningful ways. Some of their most popular workshops include:

- The Chromebook Classroom
- Formative Assessment in the Classroom
- Google Apps 1.0 (beginner / intermediate)
- Google Apps 2.0 (intermediate / advanced)
- Contact John via email (jrsowash@sowashventures.com) and start a conversation!

GET GOOGLE CERTIFIED

Attend the Google Certification Academy to earn your Google Educator certification! Find an event near you: gEducator.com

TAKE THE ONLINE COURSE

Register for the companion web-course for this book! Learn directly from John Sowash and master your Chromebook! Learn more - chrmbook.com

ATTEND THE CONFERENCE

Attend the annual summer Chrome Lab conference in Michigan! Explore the latest trends and applications for Chromebooks in the Classroom. Learn more at Chrome-lab.com.

Resources

BLOGS

Official Chrome Blog - chrome.googleblog.com

The Electric Educator (John's Blog) - ElectricEducator.com

Chrome Story: chromestory.com

Chrome Unboxed: chromeunboxed.com

OMG Chrome: omgchrome.com

TECHNICAL RESOURCES

Google Education IT Resource guide: chrm.tech/66

Attend the Google Admin Bootcamp: chrm.tech/67

Google Apps Administrators community - chrm.tech/7x

SUPPLIER RESOURCES

Device Suppliers

Acer Corporation - Tim Bigham (tim.Bigham@acer.com) acer.com

CDW - Daniel Janer (danjane@cdw.com) cdw.com

Sehi Computers - John Laswell (john@sehi.com) sehi.com

Cases

Higher Ground Gear - Todd Maddock (todd@hggear.com) hggear.com

Software

Squirrels, LLC - Jessica Chevalier (jessica@airsquirrels.com) airsquirrels.com

Pear Deck - Anthony Showalter (anthonys@peardeck.com) peardeck.com

List of Contributors

More than 60 educators from around the country contributed content and feedback for this book.

Section Editors, 2015 Chrome Lab Conference

Justin Cowen
Meghan Daniel
Eric Griffith
Meredith Nickerson
Kevin Powers

Content Contributors and Reviewers

Kym Anderson	Steve Durant	Ria Megnin
Sarah Andersen	Justin Ellsworth	Carrie Moeggenberg
Martha Ball	Steven Gale	Herb Morelock
Mary Hankins	Kim Goffee	Debra Nichols
Dave Bast	Ronald Gorbutt	Tonya Nugent
Matthew Bell	Lindsay Grady	Ashlie O'Connor
Kevin Brookhouser	Mary Hankins	Lindsay Pesonen
Danelle Brostrom	Kimberly Jasper	David Ray
Bridgett Buehrley	Valerie Johnson	Denise Rednour
Heather Carni	Erin Jones	John Sankey
McKenzie Chappell	Jim Kelly	David Schmitke
Karen Chichester	Chad Lawver	Marlene Scott
Jennifer Collins	Sam Lippert	Dakota Smith
Rick Cook	Kristina Mahaney	De'Nae Streeter
Jeff Crockett	Shawn Maison	Jennifer Wallace
Lynette Daig	Gina Matley	Amy Weston
Wheatley Davis	Matt McCullough	Ashlea Weston

About the Author

John Sowash is an experienced educator and former school administrator who challenges educators to lead their classrooms with creativity and vision. John leverages his experience as a classroom teacher and school administrator as the foundation for the training he provides for schools and universities around the world. He has a bunch of Google Certifications (all of them!) and is the founder of the Google Certification Academy and Admin Bootcamps (gEducator.com), which take place around the country.

John and his wife Nellie are serial entrepreneurs who have dabbled in retail, manufacturing, e-commerce, service, and now, the book industry. John, Nellie, and their five children live in southeast Michigan.

You can connect with John via his blog (electriceducator.com) or Twitter (@jrsowash).